Praise for America's First Great Eclipse

"This page-turning history of astronomy in the Wild West not only documents scientists understanding an elusive eclipse, but tells the tale of how astronomy itself began to change and provides glimpses of the west as it was. Loved it ... literally couldn't put it down until I finished!"

— David Lee Summers, Kitt Peak National Observatory, Author of *The Astronomer's Crypt*

"Loved it! A fun and engaging glimpse into an important era for the science of astronomy. As an astronomer I feel a greater connection to my roots and am even more excited about 2017's Great American Eclipse."

— Dimitri Klebe, PhD (Astrophysics); Founder, Pikes Peak Observatory; Co-Founder, Solmirus Corporation

"A unique history of a once-in-a-lifetime experience at high altitude. Ruskin paints a vivid picture of nineteenth-century astronomers whose initial exploits into the Rocky Mountains continue to impact modern astronomy."

— Kevin Ikenberry, US Army (Ret.), Former Manager of US Space Camp, Author of *Sleeper Protocol*

"This unique history succeeds in placing its readers in the time of an early American eclipse. Readers, be they professional or of the general public, should enjoy the experience and better anticipate any such upcoming events. A fun read!"

— Stella Cottam, PhD, Author of *Eclipses, Transits, and Comets of the Nineteenth Century: How America's Perception of the Skies Changed*

AMERICA'S FIRST GREAT ECLIPSE

How Scientists, Tourists, and the Rocky Mountain Eclipse of 1878 Changed Astronomy Forever

AMERICA'S FIRST GREAT ECLIPSE

How Scientists, Tourists, and the Rocky Mountain Eclipse of 1878 Changed Astronomy Forever

STEVE RUSKIN

Alpine Alchemy Press

America's First Great Eclipse

ISBN-13: 978-1546456285
ISBN-10: 1546456287

Cover Detail: Spectres by S.P. Langley,
New Astronomy (1888)

Cover Design by Roberto Calas, Ravenscar Designs
www.robertocalas.com

Book Design by RuneWright, LLC
www.RuneWright.com

Editing by Shelley Holloway, Holloway House
www.hollowayhouse.me

Published by
Alpine Alchemy Press

Trade Paperback Edition 2017
Printed in the USA

Contents

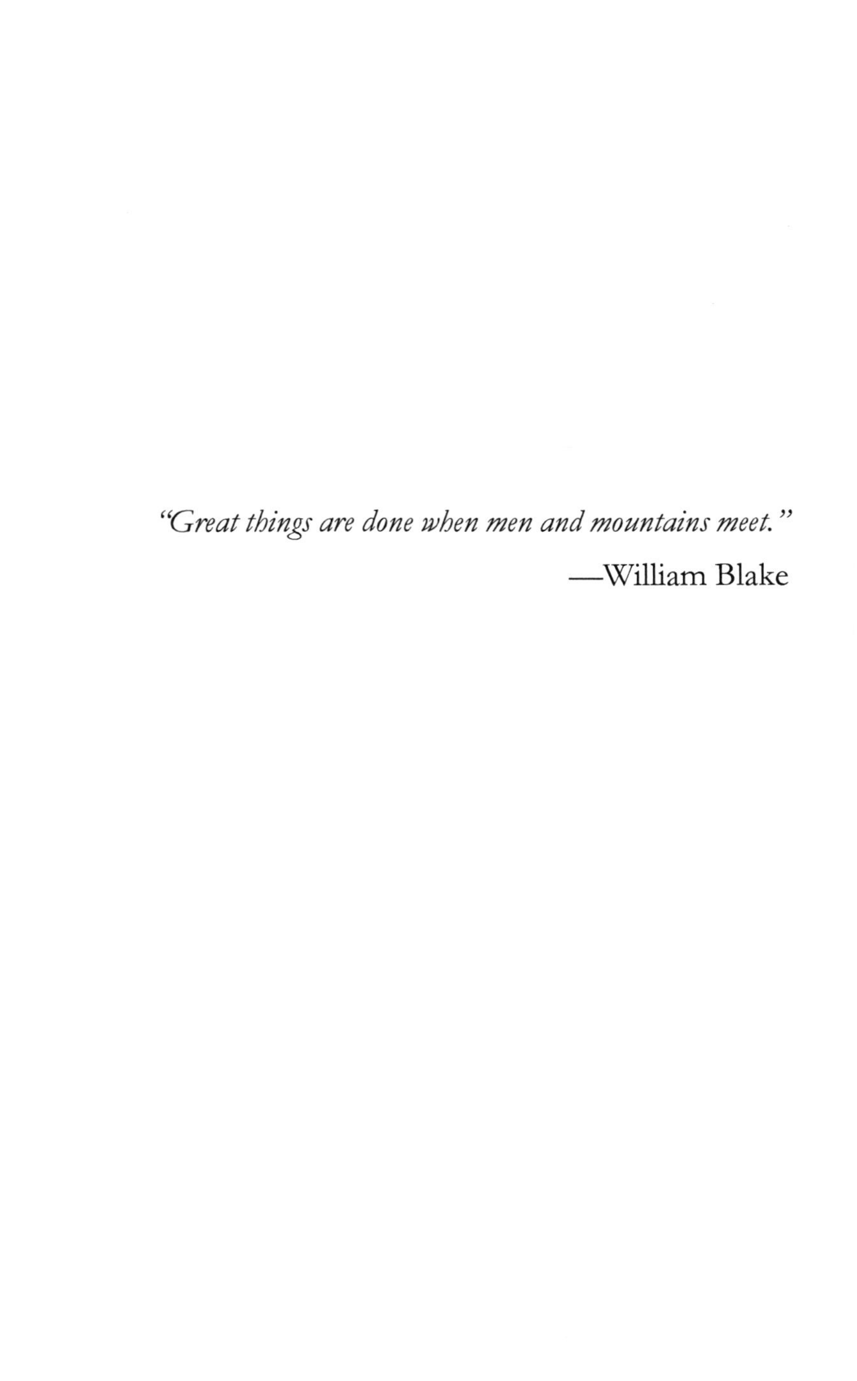

"Great things are done when men and mountains meet."

—William Blake

Sketch of the eclipse by Cleveland Abbe, Pikes Peak, Colorado, July 29, 1878

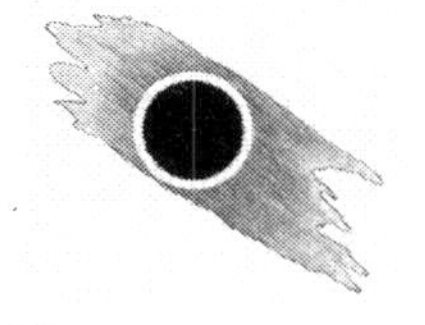

PREFACE

In 2008, I wrote a short article for the magazine *Colorado Heritage* titled "Among the Favored Mortals of Earth: The Press, State Pride, and the Great Eclipse of 1878." I am a historian of science, an amateur astronomer, and a Colorado native. So the total eclipse of 1878 is of great interest to me. I've given public lectures and talks at history conferences around the world about the eclipse. Whether my listeners were academics or just people with an interest in the history of astronomy, they always responded to the 1878 eclipse with enthusiasm, especially when they learned that Americans in the 1870s were just as excited about an eclipse as we are today, and like us, were willing to travel thousands of miles to see it.

As I write this, America is eagerly anticipating another total solar eclipse on August 21, 2017. As we get closer to that date, curiosity about previous total solar eclipses, and the way astronomers and ordinary people observed them, has only increased. To help satisfy this curiosity, I decided to compile my research and make it available in the form of this short history of the July 29, 1878 eclipse. I've kept it brief, able to be read cover-to-cover in an evening or on a plane ride. I also wrote it from the perspective of the residents of the American West, who hosted the many astronomers and tourists who flocked to the Rocky Mountains that summer, nearly one hundred fifty years ago. Back then, like now, the words of Maria Mitchell—America's first female professor of astronomy—hold true:

The entire country has been "roused to a knowledge of the coming eclipse."

PART I

PREPARATIONS

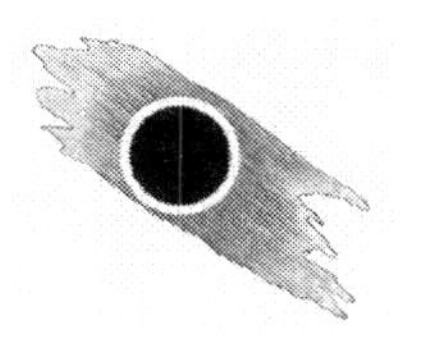

1. Looking Forward, Looking Back

A total solar eclipse is a breathtaking sight. It lasts a few hours as the moon slowly crawls across the face of the Sun, obscuring it bit by bit. But the most spectacular part of the eclipse is called *totality*: the brief period during which the Sun is completely concealed by the moon. Totality lasts only a few minutes, but its effects are dramatic. The temperature plummets as the Sun disappears, stars and planets become visible in the middle of the day, and on the surface of the Earth, a giant shadow up to one hundred and fifty miles wide sweeps across the land at speeds approaching two thousand miles per hour.

Total solar eclipses have long been a source of fascination, even before astronomy became an advanced science. By the end of the nineteenth century, however, astronomers were using those few brief moments of totality to analyze the solar corona—the Sun's gaseous "atmosphere," which was only visible during a total solar eclipse. They did so with new instruments and techniques, dissecting the Sun's light in order to better understand what elements the Sun was made of (or to determine whether the corona was even part of the Sun at all). In this, they were doing something completely new in the history of astronomy: investigating the physical nature of the Sun, and trying to figure out how its energy made possible "the Earth and our own daily lives on it."

For these reasons and more, the total solar eclipse of July 29, 1878—which occurred over the Rocky Mountains—proved to be one of the most important events in nineteenth-century astronomy. It was also an important cultural event for the United States, especially for residents of the American West.

From an historical perspective, the 1878 eclipse makes for fascinating reading. Astronomers from Europe and America traveled to the Rockies, setting up camps along the track of the moon's shadow (the "path of totality"), from Wyoming, through Colorado, and down into Texas. Thousands of tourists also came, filling hotels and boarding houses, then spilling out into the streets and nearby parks, camping in tents and even wrapping themselves in blankets and sleeping on sidewalks or in livery stables. One hotel, as you'll see, converted its billiard tables into makeshift beds for desperate eclipse tourists.

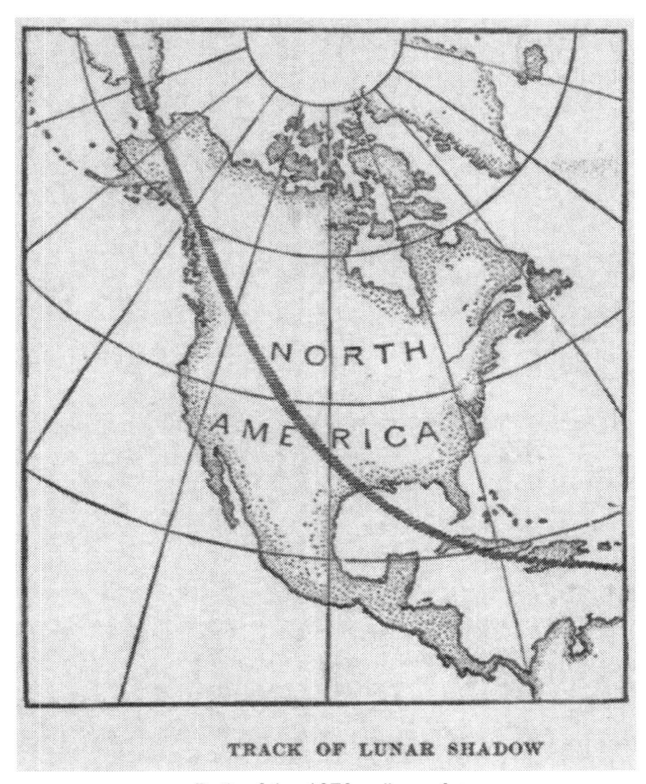

Path of the 1878 eclipse, from
Samuel Langley's New Astronomy (1888).

From the perspective of the western pioneers, the eclipse was a thrilling event. When the hordes of scientists and tourists overran their new towns and cities (especially in Colorado, which had just achieved statehood in 1876), those residents realized that the eyes of the world were suddenly turned upon their hardscrabble frontier. It was a chance to prove that even though they were still in the midst of building a new society out of an unforgiving environment, they were civilized enough to put on a good eclipse.

Non-stop press coverage put the Rocky Mountain region in the spotlight. Could the rough-and-tumble frontier host a scientific event as significant as a total solar eclipse? Would the visiting astronomers successfully make their observations, and would they have the support they needed? And, most important of all, would the weather cooperate? Even though the weather was not something anyone could control, the press, in particular in Colorado, staked their pride on the skies being clear for the eclipse, as if it were somehow the responsibility of the residents of the Rockies to sweep away the clouds above them.

Colorado was the most populous and connected of the western states and territories in 1878. Wyoming, though still a territory at the time of the eclipse, hosted a number of important eclipse expeditions because the transcontinental railroad ran right through it. In contrast, Montana was considered too dangerous—due to "Indian troubles"—to host any eclipse watchers. In Texas, the Dallas and Ft. Worth areas received only a handful of astronomers.

Most of the professional astronomers went to Colorado and Wyoming. And it seems safe to say that the overwhelming majority of eclipse tourists also went to Colorado, which—between its capital, Denver, and its largest resort town, Colorado Springs—could best accommodate them. A few of Colorado's scattered mountain towns, built largely around the mining industry, saw their populations balloon overnight, as many astronomers and tourists hoped that observing from higher elevations would give them a better view.

The eclipse that will occur on August 21, 2017 will be the first total solar eclipse visible in the contiguous forty-eight United States since 1979. It will bisect the country from Oregon to South Carolina. Not surprisingly, it is being called the "Great" American eclipse. However, it is not the first eclipse to be called "great" and actually deserve the title—that honor belongs to the 1878 eclipse.

An eclipse over the Rocky Mountains, a region most Americans knew about only in stories, gave the entire country a sense of just how great the United States had become. Although America had already witnessed many solar eclipses (the eclipse of 1869 was also occasionally described as "great"), the 1878 eclipse was the first to be a truly national event, one in which hundreds of astronomers, thousands of tourists, and tens of thousands of frontier residents would participate. The rest of the country, far from the path of totality, would wait with bated breath for the results, while the national press rightly dubbed it America's "great eclipse."

In telling the story of America's first "great eclipse," I draw heavily on records from the 1870s and 1880s, including newspaper articles, scientific reports, and other sources. I quote extensively from them, because hearing from those who saw the eclipse back then helps us relive the excitement they felt, even though we are separated from them by almost a hundred and fifty years. I hope to show that the enthusiasm we have for an eclipse today is the same as theirs was all those years ago: the emotion of pure awe that comes from witnessing one of nature's grandest spectacles.

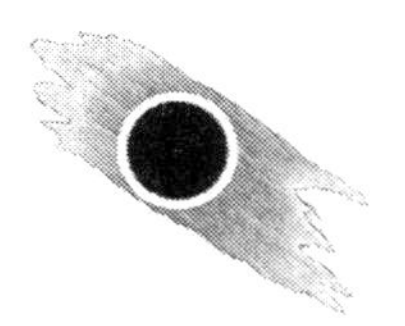

2. Each Astronomer Can Have a Peak to Himself

Total solar eclipses are spectacular events, and wherever they occur, they cause a sensation. In the nineteenth century, astronomers chased solar eclipses all over the world: to Asia, Africa, and the middle of the oceans. When it was publicized that such an eclipse would occur over the Rocky Mountains on July 29, 1878, residents of the American West were thrilled. Especially in Colorado. Just two years old (it had achieved statehood in 1876, earning its nickname "The Centennial State"), Colorado would have a chance to showcase itself while renowned scientists and hordes of tourists flocked to its mountains and cities.

In 1876, a small news item in a Denver paper may have been the first indication that in two years, an awesome natural spectacle would take place in the skies above. The March 27, 1876, *Denver Daily Times* reported that in July of 1878, a total solar eclipse would pass over the state. The paper even stated—incorrectly, but with pride—that "it does not appear that [the eclipse] will be total at any other city now existing in the United States."

The general population of the Rocky Mountain region was probably not aware that they would be hosting a major scientific event until the beginning of 1878, when various newspapers began reporting on the upcoming phenomenon. On January 26, 1878, the *Silver World* of Lake City, Colorado, noted that the eclipse would occur over Colorado in July. And by March, word was beginning to spread. The *Denver Daily Tribune* ran a short piece noting the advantage the state would afford observers:

"Colorado will have a special attraction the coming summer in the eclipse that has been appointed for July 29. It ought to attract hither large numbers of astronomers. We don't know a great deal about eclipses; but should suppose there would be special advantages in observing them from high altitudes. And Colorado will afford as high altitudes for observing as the most exacting astronomers could ask, and enough of them to accommodate all the astronomers in the country. Long's, Gray's, Pike's and James' peaks will all be included in [the eclipse path]. And there are within the same belt innumerable other peaks less noted but nearly as high. Each one of the rival astronomers can have a peak to himself."

By April, newspapers across the Rockies were publishing the exact time the eclipse would begin, or noting, as did the Laramie, Wyoming *Daily Sentinel*, that parts for one telescope had already arrived.

But could the astronomers afford to come? Eclipse expeditions were not cheap, and the United States government had still not appropriated the funds necessary to send astronomers out west.

The *Denver Daily Tribune* of April 20, 1878 complained loudly that the coming eclipse—perhaps one of the most significant scientific events in the country's history—was being overlooked in Washington because it would occur in the Rockies. The indignant *Tribune* chided Congress, suggesting the lawmakers would rather ogle women than help the scientific community observe an eclipse:

"Congress has as yet made no appropriation for the observation of the total eclipse of the Sun ...this seems to be another instance of unjust discrimination against the far West. ...If the approaching total eclipse had been appointed for the White Mountains, for Boston, for New Orleans, for the peaks of the Virginia mountains, for Philadelphia, for Chicago ... Congress would have treated it with proper respect and due consideration ... and would have made ample provision for its observation. In fact, Congressmen would no more have allowed it to go unobserved that they do the handsome female lobbyists that visit their hall. But as the eclipse has been appointed for Pike's, Long's and Gray's peaks, and for Denver, Colorado Springs and Greeley, Congress does not seem to care whether any observations are made of it."

Within a few months, however, the Colorado press happily reported that Congress *did* in fact care about them. The *Colorado Weekly Chieftain* of July 11, 1878 noted: "Congress has appropriated $8,000" for eclipse observations. The money was given to the United States Naval Observatory to administer and fund six different eclipse expeditions to the Rocky Mountains. The *Chieftain* went on to describe the composition of the various expeditions—their leaders, the astronomers who would be making observations, and the locations from which the eclipse would be observed. These included Creston and Rawlins in Wyoming, and Denver, Central City, and the summit of Pikes Peak in Colorado, among others.

Signal Station on Pikes Peak, c. 1876.
Courtesy Pikes Peak Library District Special Collections.

Although an important expedition, which included Simon Newcomb (then director of the Navy's Nautical Almanac Office) and Thomas Edison, went to Rawlins, and other astronomers went to Texas, Colorado got the lion's share of eclipse expeditions. One group made up of observers from the US Naval Observatory, Johns Hopkins University, and West Point went to Central City to observe the eclipse from the roof of the Teller Hotel. Astronomers from New York and St. Louis went to Idaho Springs.

Charles Young, a proponent of high-altitude astronomical observations, led the Princeton University expedition that observed at Cherry Creek, near Denver. The young astronomer William Pickering, brother of Harvard College Observatory director Edward Pickering, also observed near Denver, as did Maria Mitchell, the trailblazing astronomy professor from Vassar College.

Samuel Pierpont Langley, of the Allegheny Observatory in Pennsylvania, decided to scale Pikes Peak, near Colorado Springs, with his brother John Langley and the meteorologist Cleveland Abbe. Asaph Hall, an astronomer who had just discovered the moons of Mars in 1877, took his expedition to La Junta, in the southeast part of Colorado.

The expeditions usually included a support crew, to set up camp and maintain their delicate instruments, and aid in other important technical aspects of the observations. These expeditions will be discussed in more detail later. Suffice it to say, the residents of Colorado and Wyoming were abuzz with excitement. The whole world would observe the eclipse through their eyes. And as with any major event, it was bound to be widely reported in the local and national press. Not only would the eclipse shine a spotlight on the American frontier, but it would also bring the region scientific notoriety.

It was the first time that such a large number of scientists visited the frontier at one time. And, significantly, it was also the first time that so many astronomers observed together at higher elevations.

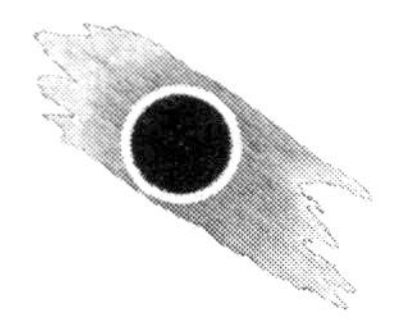

3. ASTRONOMY AT ALTITUDE

Before I discuss the 1878 eclipse, a little scientific background is necessary. (If that doesn't interest you, feel free to skip to the next chapter.) The end of the nineteenth century was the period of the "new astronomy," an era of modernization for this very old science. From ancient times to the middle of the nineteenth century, astronomy was primarily a positional science. That is, astronomers were concerned with the positions of objects, like stars, nebulae, planets, and the moons of Jupiter. Knowing these positions, and being able to calculation those objects' future positions, allowed astronomers to create accurate tables and charts to help navigators determine their location as they sailed the seas.

Questions about what exactly the Sun, stars, nebulae, or even the planets were made of had only occupied the minds of relatively few astronomers prior to the nineteenth century. Up until the middle ages, it was assumed that the heavens were made of very different stuff than what we found on Earth, and operated under different rules. Nicholas Copernicus, Galileo Galilei, Johannes Kepler, and Isaac Newton gave us reason to believe that the heavens and Earth were parts of the same whole, obeying the same universal laws.

It wasn't until the middle of the nineteenth century, notes historian Michael Crowe, that "astronomers gained access to a number of new instrumental techniques, in particular, photography and spectro-scopy." They recognized that these tools would be a great help in investigating distant celestial objects, allowing astronomers to suddenly become astro-physicists and astrochemists. The English scientist William Huggins was one of the first to use spectroscopy to "unravel starlight," as historian Barbara Becker puts it. Huggins used a spectroscope—a specialized device with a prism—on the light of distant stars, to show that those stars were made of the same chemical elements we had here on Earth. "No longer," observed historians Stella Cottam and Wayne Orchiston, "would astronomy be considered only a tool for navigation and time-keeping."

The "new" astronomy wanted to extend this kind of investigation; to analyze the composition of the Sun and other celestial objects. Astronomers wanted to know not just where those objects were, but *what* they were made of. And so they started to apply new discoveries in physics, chemistry, and other fields to the heavens. They used more than just telescopes. They employed cameras, spectroscopes, and other instruments to investigate objects beyond Earth. And the Sun was their first target.

Samuel Langley was an astronomer, physicist, and aviation pioneer, for whom Langley Air Force Base is named. In 1884, he described the new astronomy as an extension of terrestrial physics into the celestial realm. "Within a comparatively few years, a new branch of astronomy has arisen, which studies Sun, moon, and stars for what they are in themselves. Its study of the Sun, beginning with its external features, led to the further inquiry as to what it was made of, and then to finding the unexpected relations which it bore to the Earth and our own daily lives on it. This new branch of inquiry is sometimes called Celestial Physics, sometimes Solar Physics, and is sometimes … referred to as the New Astronomy."

But the eclipse of 1878 offered more than just a chance to apply the techniques of the new astronomy to the Sun during a total solar eclipse. It also offered the opportunity to experiment with making astronomical observations at higher elevations.

Since the twentieth century, astronomers have taken for granted that observations made at higher elevations are—all other conditions being equal—better than those made at lower elevations. High-altitude observations provide less atmospheric interference and therefore superior results. Today, most of the world's largest telescopes are set high on mountaintops (or at least well above sea level), and some, like the orbiting Hubble telescope, are outside our atmosphere altogether. But astronomers did not always feel this way about high-altitude observing. The 1878 eclipse was the first time a large group of astronomers tested the advantages of taking their telescopes and other instruments to higher elevations.

Prior to 1878, the advantage of observing from higher elevations was still a matter of speculation. In 1896, the bespectacled, owlish astronomer Edward Singleton Holden (who observed the 1878 eclipse from a hotel rooftop in Central City, Colorado) remarked that

Astronomers prepare to observe the eclipse from the roof of the Teller House hotel in Central City, Colorado (elevation 8,500 feet). Courtesy U.S. Naval Observatory Library.

determining what made "conditions suitable for *astronomical* work at high levels" was a process of trial-and-error. It was, however, the "eclipse-expeditions of 1878 to the Rocky Mountain region" that first "familiarized many observers with the question."

That is one reason the 1878 eclipse was so eagerly anticipated by astronomers. The fact that the path of the eclipse lay directly over the Rocky Mountains was simply a fortuitous accident of nature. But it was an accident that presented a tremendous scientific opportunity.

The history of astronomy includes centuries of speculation on the possible advantages of mountain-top observatories. In his 1704 book *Opticks*, Isaac Newton was perhaps the first to suggest astronomy would benefit from higher altitudes. "The Air through which we look upon the Stars," he wrote, "is in a perpetual Tremor. Telescopes cannot be so formed as to take away that confusion of the Rays which arises from the Tremors of the Atmosphere. The only Remedy is a most serene and quiet Air, such as may perhaps be found on the tops of the highest Mountains."

In the twenty years prior to the 1878 eclipse, two expeditions tested this hypothesis. In 1856, Charles Piazzi Smyth, the Astronomer Royal for Scotland, went to the volcanic island of Tenerife in the Atlantic Ocean, and in 1872 the US Coast Survey sent Charles Young to the high plains of Wyoming.

During that summer of 1856, Piazzi Smyth set up his telescope at two different elevations (approximately nine-thousand and eleven-thousand feet). Despite struggling with wind, clouds, dust, and rain, he nevertheless "deduced an improvement of astronomical vision with height." He compared observations he made in Edinburgh, where he could see stars up to the tenth magnitude, to those on Tenerife, where he could see stars up to the fourteenth magnitude (the higher the order of magnitude, the fainter the star). He concluded that higher elevations indeed gave tremendous advantages to astronomers.

Holden, however, thought Smyth's widely-circulated report overstated the case for high-altitude astronomy. "There is no doubt that [Smyth's] expedition served to attract general attention to the matter of choosing suitable sites for observatories; and also to spread the idea that *all* mountain-stations possessed striking advantages."

It wasn't until Charles Young went to Wyoming in 1872 that the advantages of higher elevations were investigated from an American perspective. In 1868, the American Association for the Advancement of Science passed a resolution arguing (in Holden's words) for the "importance of an Astronomical Observatory at some point on the Pacific Railroad between Nebraska and the Pacific Coast, and at as high an altitude as possible, where clearness of the atmosphere and the great number of cloudless days would ensure remarkable and unsurpassed opportunities for astronomical observations."

The AAAS passed another similar resolution in 1870, finally resulting in an appropriation from Congress for the project. The money was administered by the Coast Survey. Under this appropriation, Young was sent to Sherman, Wyoming. He operated out of a small wooden shack at an elevation of eight thousand feet, near the highest point of the original Union Pacific segment of the first Transcontinental Railroad. With an escort of soldiers from the Army (the area was still considered to be "Indian country"), it was reported that Young's "telescope and spectroscopic apparatus were the largest and most powerful ever mounted at so high an altitude."

Young was widely regarded as one of America's most accomplished solar astronomers and spectroscopists. In 1872, he was in Wyoming primarily to make observations of the Sun: sunspots, the solar chromosphere, and solar storms. But the clear nights at that elevation provided him an opportunity to observe the stars and planets as well. He stated unequivocally that the high western plains enabled observations far superior to those he'd ever made at sea level:

"When the sky was clear, it was beautifully so. At night, hundreds of small stars, which I never saw at home, came out distinctly … the background being far more dark, and the images cleaner and sharper than I had even seen them. I have no hesitation in affirming that I never was able to see [the stars at sea level] with anything like the perfection attainable here."

This was high praise, and it whetted the appetites of other astronomers eager to observe in the American West. Young made the remarkable claim that the altitude and atmosphere of Wyoming actually enhanced the power of his telescope: "It is my deliberate opinion that at Sherman, my 9.4 [inch] object-glass was just about equal to a 12-inch glass at the sea-level." In other words, a smaller telescope at higher elevations was equal to a much bigger telescope at lower elevations.

But a year later, in 1873, Holden, then with the US Naval Observatory, made similar observations in Colorado. He was trying to determine a possible site for the Naval Observatory's brand new, forty-foot-long, $50,000 telescope—at the time, the world's largest refracting telescope. He disagreed with Young, and Holden's government position gave him considerable weight. "I made a stay in Colorado and reported (adversely) on the placing of a great telescope in any of the stations." And Holden, it is important to note, became the first director of California's Lick Observatory in 1888. The Lick is a mountaintop observatory, the first in the world in fact, but it sits at the fairly moderate elevation of 4,209 feet. So he was not opposed to high-altitude astronomy on principle, but rather on matters of practicality.

So, prior to 1878, there was no general consensus on the advantage of high altitudes for astronomical observations. Consensus would require a large number of astronomers, observing together at a high altitude, and then a comparison of results. And that opportunity is exactly what the 1878 eclipse provided. Yet even though the astronomers would observe the eclipse in July—midsummer in North America—they still had to prepare for the harsh conditions of their temporary high-altitude observatories.

4. Blankets and Brandy

The eclipse of 1878 was the first time a large number of professional astronomers observed simultaneously at high altitude. Prior to the end of the nineteenth century, most of the world's astronomical observatories and stations had been at or near sea level, largely because of astronomy's crucial job of providing national navies with accurate celestial charts for navigation.

Astronomers who went west in 1878 camped within the narrow band of the eclipse path, a line defined by the moon's shadow during the eclipse. That line (as noted earlier, also known as the shadow path, or line of totality) was roughly one hundred and thirty miles wide, but thousands of miles long. Viewing the eclipse from within that shadow path was the only way to see the total eclipse; outside the shadow, one could only see a partial eclipse.

The width of any shadow path varies from eclipse to eclipse, but is usually between one hundred, and one hundred and fifty miles. However, the length of the path runs for thousands of miles, as the Earth rotates beneath the eclipsed Sun. And the shadow's speed is phenomenal. Depending on the time of day, and the location of the observer, the moon's shadow can approach speeds of two thousand miles per hour.

The astronomers were therefore constrained in their choice of location by the width of the moon's shadow. But staying within that narrow band, they spread out along a thousand mile length between Wyoming and Texas. Those to the north, in Wyoming, would see the eclipse first, minutes before those in Colorado.

The expedition to the highest point of observation was led by Samuel Langley. He, accompanied by his brother John and the meteorologist Cleveland Abbe, were atop Pikes Peak (14,115 feet). They had absolutely no idea what they were in for, as Langley later reported.

Even in summer, especially at the higher elevations of the Rocky Mountains, snow and violent storms were possible, and the nights could be frigid. Few astronomers had ever observed at any altitude above sea level, so there was little in the way of useful knowledge about how to prepare for such conditions. In fact, the advice the astronomers shared amongst themselves was laughably basic: dress warmly, bring blankets, and don't forget the alcohol.

Prior to heading west, Simon Newcomb, director of the Nautical Almanac Office, wrote to the English astronomer Sir J. Norman Lockyer: "Bring heavy blankets or rugs, or both, as the nights are extremely cold … with a flask of brandy in case the water is unwholesome I think health and comfort can be assured."

The comforts of home were important, as were the conveniences of civilization. Henry Draper, who observed the eclipse near Rawlins, Wyoming, chose that location in part because it had "the advantages of being supplied with water from the granite of the Cherokee mountain, and of having a repair shop, where mechanical work could be done." Although Rawlins was relatively small—a freight stop along the Transcontinental Railroad—Lockyer, who also came to Rawlins, called it a "thriving city" at seven thousand feet.

All of the astronomers came west by train. Railroad companies were happy to do their part for science: astronomers from Europe and the United States were given half price fare from the east coast to Denver (via Chicago or St Louis) by the Pennsylvania Railroad Company. Other railroads offered similar discounts to points in Wyoming and elsewhere.

Because the astronomers brought thousands of pounds of equipment and provisions—large brass telescopes are quite heavy, even when disassembled—their observation camps were usually set up close to the railroad tracks, ideally as close to a town or small railway station as possible. The locations of their camps were essentially chosen for them by the projected path of the eclipse: wherever a railway line and the eclipse path intersected, you were likely to find one of the astronomical expeditions. Fortunately for the astronomers, the 1878 eclipse path did have a number of small towns, and even one or two cities, along it.

Denver in 1878. Courtesy Pikes Peak Library District Special Collections.

Denver, which just so happened to be right in the path, was chosen by many of the astronomers for its location and amenities. But as a group, the astronomers dared not put all their eggs in one basket. If they all had gone to Denver, and that city was overcast on July 29th, no one would see the eclipse. Spreading out along the eclipse path increased the odds that at least some of the observing parties would encounter good weather.

* * *

The late nineteenth century was a time when American astronomers, in the words of historian Stephen G. Brush, "managed to rise to world leadership in a relatively 'pure' branch of science." This fact was widely recognized by European astronomers. In anticipation of the 1878 eclipse, the English astronomer Lockyer (who was also the founder of the journal *Nature*), commented:

"There is no doubt whatever that the eclipse which will sweep over the United States next July will be observed as no eclipse has ever been observed before. The wealth of men, the wealth of instruments, and the wealth of skill in all matters astronomical, already accumulated there, makes us Old Country people almost gasp when we try to picture to ourselves what the golden age will be like there, when already they are so far ahead of us in so many particulars."

As far back as 1869, Lockyer himself anticipated going to America to view the eclipse of 1878, noting that no one, especially Americans, ever came to England to observe eclipses because "it is a spot where the Sun steadily refuses to be eclipsed." But he set expectations high for 1878, especially for American astronomers, and helped build the excitement when he wrote:

"Draper, Hall, Harkness, Holden, Langley, Newcomb, Peters, Pierce, Pickering, Rutherfurd, Trouvelot, and last, but not least, Young are the names that at once run easily off the pen to form a skeleton list, capable of considerable expansion with a little thought, when one thinks of the men who will be there. One knows too that all the enthusiasm of devoted students and all the appliances of modern science will not be lacking. So that we may be sure that not only all old methods but all possible new ones will be tried to make this year one destined to be memorable in the annals of science."

To put Lockyer's comment in perspective, understand that solar eclipse expeditions were the nineteenth-century equivalent of our modern space program. Science writer Trudy E. Bell put it this way: "Total-solar-eclipse expeditions in the last half of the 19th century were the most advanced field expeditions of any science, technically equivalent to NASA's planetary flybys or manned lunar landings a century later." What impressed Lockyer was that the Americans were willing to commit so much in terms of their money, time, and expertise to something with such a high possibility of failure. After all, as Bell notes, eclipse expeditions spent "untold dollars, months, and sweat" for just the mere possibility of "observing a unique event shorter than 450 seconds, with no second chance." Like current space programs, the costs were high, and the risks often higher.

As the eclipse approached, astronomers arrived in the American West by the dozens. They had prepared as well as they could to observe in the unforgiving conditions of the mountains and the high plains. But for all their preparations, they were still at the mercy of the Rocky Mountains and the fickle weather systems that tumbled over them.

They knew good weather would be helpful as they acclimated to their campsites and set up their equipment. But it would be vital on the day of the eclipse. One wayward cloud, to say nothing of a massive thunderstorm, could block the eclipse from view, rendering months of preparation, and thousands of miles of travel, utterly worthless.

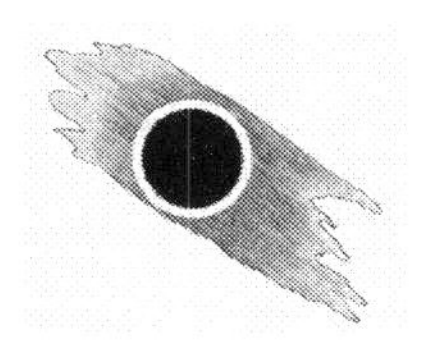

5. Prospects for a Good Day

With so many astronomers coming west for the eclipse, to say nothing of all the tourists, frontier pride was on the line. Imagine if, after all the effort and expectation, cloudy skies or a rainstorm prevented the eclipse from being observed—bad weather would spell disaster. So we can understand the anxiety behind the article that ran in the *Denver Daily Tribune* on July 14, 1878, a little over two weeks before the eclipse. Titled simply "The Eclipse and the Weather," this article reviewed weather records for the date of July 29 over the preceding six years to determine the odds of having good weather for the day of the 1878 eclipse.

"There are so many scientific parties coming to Colorado to observe the approaching eclipse that great interest is felt in the condition of the weather, upon which everything depends. Without assuming to say what the weather shall be, Sergeant Barwick, of the local signal office, has put himself to the trouble of searching the records of the office to learn what the weather has been. The Sergeant says that in comparing the years on which July 29 was favorable for observing the Sun, with the rainfall of June, he finds that the month preceding July of each year, which gave a heavy precipitation of vapor or moisture, has been followed by a favorable 29th of July."

The article then discussed the typical summer cloud cover. "The kind of clouds most prevalent during the afternoons, are cumulus and cumuli-stratus. The cumulus is a heavy, dense, and thick cloud, mostly white, except when blended with the stratus, when it becomes cumuli-stratus, having a dark base and white tops. The rainfall for June, 1878, was the heaviest ever known in that month, and it may be a good omen for the weather on July 29, 1878."

The article then proceeded to give an account of the "state of the weather" at 2:43 p.m. (the predicted time of totality in Denver) on July 29, for each year since 1872. The report was optimistic: the preceding six years produced four clear July 29ths, and only two cloudy ones. Statistically, the odds were good, and the article concluded that it was "to be hoped that we will have the average rain fall during the next week, thereby giving us better prospects for a good day." All the

Tribune could do was to "hope that Colorado will give us one of her best days on the 29th."

Transit and Eclipse.

The transit of Mercury, which occurs on the 6th of May, will begin, Denver time, at 8:12 a. m., and end at 8:45 p. p. The eclipse of the sun, occurring on July 29, 1878, will be, by Denver time, as follows:

Eclipse begins..............................2:16 p. m.
Totality begins.............................3:30 p. m.
Totality ends...............................3:33 p. m.
Eclipse ends................................4:31 p. m.

Predicted times of the beginning, totality, and end of the 1878 eclipse. Notices like this one were published in newspapers all across the Rocky Mountains. Gilpin County Evening Call, April 30, 1878.

However, in the week leading up to the eclipse, the previous years' data appeared to be of no help. Rain fell intermittently across Colorado and Wyoming, and summer thunderstorms obscured the afternoon skies. Just a few days before the eclipse, the July 27th edition of Pueblo's *Colorado Daily Chieftain* lamented, "If these thunderstorms continue every afternoon, our chances for seeing the eclipse will be small."

And yet out on the plains just east of the Rockies, the summer heat was intense. One humorous anecdote, from an anonymous Pueblo resident, described both the heat and the festive atmosphere that the eclipse was producing among the locals. That resident had received an invitation from a friend to come to La Junta, Colorado to watch the eclipse from there. The astronomer Asaph Hall, from Washington, DC, was observing in La Junta, and his expedition's presence was probably the biggest event to ever happen in the area.

"Come down," wrote the friend. "We are going to have an eclipse. All the boys are here from Washington, and we have ten barrels of beer."

The reply? "It would take something more than an eclipse and ten barrels of beer to tempt us to brave the dangers of an eighty-mile ride through the dust with the thermometer 100° in the shade."

If the locals were wary, the tourists were undaunted. As the day of the eclipse approached, trains coming from the east disgorged hundreds of visitors a day, all eager to explore the Rocky Mountains and, on the afternoon of the 29th, join the astronomers in staring up at the sky.

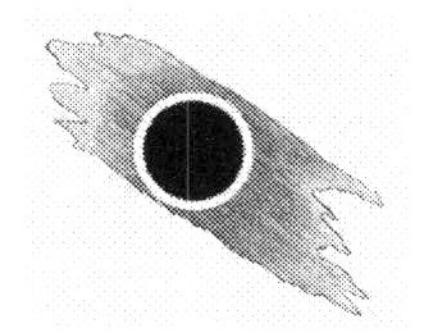

6. A Mammoth Excursion

Solar eclipses were hugely popular events in the nineteenth century, just as they are today. Regular people, especially those with the financial means, traveled to locations within an eclipse path just for the sheer spectacle of observing it, and the social cachet of being observed by others. In fact, the historian Alex Pang has referred to nineteenth-century solar eclipses as "the social event of the season," for their ability to attract large numbers of tourists alongside professional astronomers. After all, when an event is talked about for months or even years ahead of time, anyone who's anyone would want to be there. One Colorado Springs hotel is even reported to have hired an Italian band to play Beethoven during the eclipse for its posh guests, helping to make the event a cultural, as much as a scientific, one.

Colorado played host to the majority of the eclipse tourists, and Colorado Springs was a particularly popular spot. Founded in 1871, it was a resort community, with healthy air, abundant sunshine, nice hotels, and nearby mineral springs, all tucked into the foot of the Rockies beneath the looming shadow of Pikes Peak. In June, the *Chicago Times* reported on a "mammoth excursion from the [Great] lakes to the mountains," with all those tourists heading for a famous red-sandstone park just outside of Colorado Springs, called the Garden of the Gods. European tourists were making the trip as well. Colorado Springs had a large contingent of English expatriates, eventually earning it the nickname "Little London."

In the week before the eclipse, the *Colorado Weekly Chieftain* reported that "Colorado Springs, Manitou, the Garden of the Gods, and Pike's Peak are being thronged with visitors from Europe and the states." Indeed, even members of Congress were there. The *Chieftain* continued, "Many United States senators, under the lead of Senator [Henry] Teller, are already at the Manitou House, and others are expected." Tents were set up in the Garden of the Gods, where spectators could watch the eclipse for twenty-five cents.

A few miles away, in the streets of Colorado Springs, every available viewing spot was taken. "Seats upon the balconies of our best houses are all engaged, and elevated platforms are being erected upon our public square. The windows in our church steeples have been leased—those facing the eclipse at fifty cents and those facing in the opposite direction at half price."

Interest in the visiting scientists was almost as great as interest in the eclipse itself. Newspapers provided regular notices of the arrivals of astronomers from the East Coast and Europe. The *Denver Daily Tribune* reported almost daily during the month of July on the astronomers arriving in that city. As early as July 7, the paper announced that the Princeton University expedition had arrived, consisting of Charles Young and ten others, some of who were recent graduates. The paper noted that the Princeton party came "equipped for all kinds of observations that are calculated to promote interests of science," and would "go into camp near Denver."

The astronomers arrived early in order to scout out ideal observing locations, set up their camp, and prepare their instruments. The Princeton expedition was the trickle before the coming flood. As noted by the *Denver Tribune*: "This is the first of many parties who are expected to come to Colorado to view the eclipse. Reports almost every day bring information of the purpose of new scientific parties to make this their headquarters for observation." And back east, the *New York Times* reported: "Astronomers and scientists are already gathering in Colorado in large numbers to witness the coming eclipse of the Sun."

The astronomers were like celebrities in the frontier towns. They brought an aura of scientific respectability and a dose of excitement, and residents followed them around all day long. The flattered astronomers courteously offered public demonstrations of their telescopes and other equipment, but even the little details of their daily lives were under the microscope.

It was as if the frontier press employed paparazzi: every meal the astronomers ate, every hotel they checked in to, every comment they made about local observing conditions, all of it was reported in a flurry of little notices, as if the astronomers were visiting royalty. "E.S. Holden and C.F. Hastings, of the United States Navy, are visiting Denver and are registered at Charpiot's Delmonico," reported Denver's *Daily Tribune*, in case their readers wanted to stalk the astronomers at their hotel.

The *Colorado Springs Gazette* eagerly reported on the arrival of Samuel Langley's scientific instruments at the local signal-service office. The piled freight, containing a large dismantled telescope and other instruments, probably took up half the office as locals crowded the other half to watch the contents being prepared for transportation up the rocky, rough trails to the top of Pikes Peak.

Two years before the eclipse, in 1876, a small stone building had been erected on the boulder-strewn summit of Pikes Peak. The building was manned year-round by officers of the Army's signal service. At the time, it was the world's highest weather station and was connected by telegraph down to Colorado Springs. The arrival of Langley's telescope was big news for the little town. "They arrived yesterday from Washington, and came in eleven boxes. The instruments will be packed to the summit on the backs of burros, and much care will necessarily have to be taken in transporting them."

Finally, as the day of the eclipse drew nearer, the Laramie, Wyoming *Daily Sentinel* exuded excitement that "scientists from every quarter of the globe inhabited by intellectual beings are today within the limits of totality." As the Boulder, Colorado *County Courier* put it: "To those who admire genius, a good opportunity will be given of seeing the objects of their admiration, at Denver, through the large number of scientific men who will gather at that point to view the total eclipse." For the public, watching the astronomers setting up their camps and telescopes was almost as much of a spectacle as the eclipse itself promised to be.

* * *

Perhaps the most significant testaments to the cultural significance of the eclipse were the men of power and prestige who came west to participate in the viewing. Senators have already been mentioned, but there were business moguls, magistrates, and military men as well. They traveled west in luxury and style.

George Pullman, president of the famous Pullman sleeping car company, came to Colorado in his private rail car. Riding along with Pullman—relaxing in velvet-covered chairs, drinking whiskey and playing cards—were a cadre of equally impressive guests: an Indiana senator, a district judge, a municipal judge from Cincinnati, and a US Army General. A California senator joined the party in Denver. In Wyoming, territorial Governor John W. Hoyt made his way from Cheyenne to view the eclipse in Rawlins. (Hoyt actually managed to draw a very competent sketch of the Sun's corona during totality.) Along with the astronomers, the presence of political, judicial, and military leaders brought a certain sense of gravitas to the event.

As frontier towns prepared to observe the eclipse, competition over which of them was best situated along the shadow path became fierce. Colorado's newspapers were self-conscious; they were under a microscope, and the eyes of bigger newspapers like the *New York Times* and *Washington Post* were on them. Stories printed locally were often reprinted in those eastern newspapers, and each local publication touted its city's relative merits and challenged the reports of their rivals. Not long before the eclipse, the *Denver News* fired the first volley by warning eclipse tourists away from the higher elevations in July, because "the rainy season begins in the mountains about that time."

The *Georgetown Courier* of July 25, 1878, quickly defended its mountain community, retorting: "As the *News* seems to know all about it, will it be kind enough to inform us all about the 'rainy season,' for we freely confess that we are unable to see any system whatever to the rain storms of the Rocky Mountains. …We would like to know upon what authority the *News* tells its readers that they had better stay in Denver."

In the end, Georgetown seems to have won bragging rights in this little tussle. A few weeks after the eclipse, a spectacular view of the event as seen from Argentine Pass (high above Georgetown) graced the front cover of *Harper's Weekly: A Journal of Civilization*. It was one of the most memorable and widely distributed images of the eclipse, and Denver was undoubtedly envious. But that was still to come. First, Colorado had to find somewhere to house all those tourists.

HARPER'S WEEKLY.

A JOURNAL OF CIVILIZATION

NEW YORK, SATURDAY, AUGUST

Cover of Harper's Weekly, August 24, 1878.

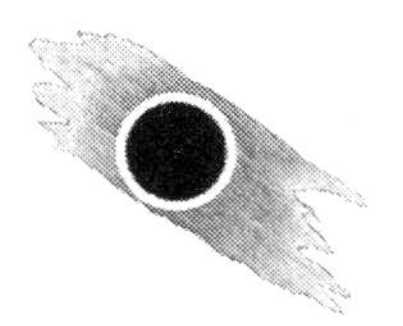

7. Crowded with Tourists

Tourism was important to the western frontier, and the 1878 eclipse was about the best natural advertising residents of the Rocky Mountains could have hoped for. It was a huge boon for the cities and towns in its path, bringing in revenue for the local businesses, and also showcasing the region. In a 1972 article in *Sky and Telescope*, astronomer John Eddy commented: "In the *American Ephemeris and Nautical Almanac* for 1878, the mountains and towns given in the predictions for the coming eclipse had read like a tour guide of the Rockies! The new transcontinental railroad, completed only nine years before, made western travel easy."

Western railroads like the Denver & Rio Grande took passengers throughout the mountains on sightseeing excursions before, during, and after the eclipse. Once easterners could travel in style, they were drawn to the majestic beauty of the Rockies, giving the new towns and cities a chance to convert tourists into permanent residents.

A few days before the eclipse, a Denver paper commented that the event would be "watched by thousands of residents of Denver and vicinity, besides no small number of tourists, to whom the eclipse was the turning point in selecting a summer tour."

In fact, tourists started arriving more than two weeks before the eclipse, having planned their summer vacations in the Rockies around it. On July 13, Georgetown's *Colorado Miner* commented, "part of the eclipse sight-seekers have already arrived, and gone into camp in a very charming part of the city's suburbs." The report went on to note, with evident humor, that when the astronomer Maria Mitchell inquired by mail whether or not there was "any level space near Denver from which she could make her observations," she was told "between Denver and Kansas City . . . there might be room sufficient."

On July 22nd, the word out of Colorado Springs was: "our city, Manitou, the Garden of the Gods, and Pike's Peak are being thronged with visitors from Europe and the states who have come that they may be in near proximity to the Sun during the passage of the moon across its track on the 29th." And the *Denver Daily Tribune* of July 25th predicted the Rocky Mountains would "abound in astronomical truth seekers, non-professional as well as professional. Every man, woman, and child may be expected to have an eye to the Sun's interest and conduct." Meanwhile, local residents not within the eclipse path were making their plans for "eclipse excursions to Denver and other points of observation" on the big day.

Garden of the Gods near Colorado Springs, where everyone from US senators to tourists watched the eclipse (Harper's Weekly, July 19, 1879). Courtesy Pikes Peak Library District Special Collections.

The *Colorado Weekly Chieftain* reported that railway travel "from the east is very heavy. The Atchison, Topeka, and Santa Fe train brought in one hundred and seven passengers on Saturday [July 13] ...the train having two sleeping coaches attached." With over a week to go, the July 20th *Boulder County Courier* reported: "Tourists are coming into Denver so thick that they can hardly find hotel accommodations. It looks as though by the time the eclipse gets around there will be no longer room in-doors, and that tents will be the only resort of the more tardy visitors."

Mitchell, who came to Denver with a team of female students, commented as they traveled by train from Boston: "It had been evident that the country was roused to a knowledge of the coming eclipse; we overheard remarks about it; small telescopes travelled with us, and our landlord at Kansas City, when I asked him to take care of a chronometer, said he had taken care of fifty of them in the previous fortnight."

In the days leading up to the eclipse, hotels in the vicinity of the eclipse path were full to overflowing. "Denver hotels are crowded with tourists," reported the July 19th *Laramie Daily Sentinel*. Bed capacity had been reached and exceeded—some hotels were setting up cots and renting out any other available space, including the tops of billiard tables. Reported one newspaper: "In Denver are found such entries as these in the registers of the hotels: J.E. Landholm, parlor; F.R.S. Wright, cot, dining room; H.B. Jenkins, billiard table."

In Colorado Springs, according to the *Gazette*: "One of our popular landlords visited all the livery stables and engaged all the vacant stalls." The *Gazette* concluded that, despite the fully booked hotels, no one need be turned away; if nothing else, guests would be given a "Tucson blanket" (presumably a Native American rug) and allowed to sleep outside. And in Rawlins, Wyoming, the great Thomas Edison—perhaps the biggest scientific celebrity to come west to view the eclipse—was lucky to get one of the last available rooms in the town's only hotel.

Although it is difficult to know exactly how many tourists came, we can guess based on the number of hotel rooms they filled, at least in the two biggest cities in Colorado. According to the 1878 Denver City Directory, Denver had thirteen hotels and thirty-seven boarding houses. The 1878 Colorado State Business Directory indicated Denver had forty-four hotels (which presumably includes boarding houses). Thus, Denver had somewhere between forty-four and fifty lodging options for tourists and astronomers.

Similarly, the 1879 Colorado Springs Business Directory listed eight hotels and thirteen boarding houses, while the 1878 Colorado State Business Directory stated the Colorado Springs area had a total of fifteen hotels and boarding houses (three boarding houses and four hotels in Colorado Springs, seven hotels in nearby Manitou Springs, and one hotel in neighboring Colorado City). So Colorado Springs had approximately fifteen to twenty-one places of accommodation for eclipse travelers. These numbers obviously vary based on the source, and most certainly do not account for rooms available in private residences, or tents pitched in private yards, or repurposed horse stables. However, if the newspaper accounts were not wildly exaggerated, then it would have, indeed, taken a good number of tourists to fill these hotels and boarding houses to capacity—and beyond.

But how many tourists would that be? We have one clue from a publication by the geographer William Wyckoff, who noted that by the end of the 1870s, the Colorado Springs–Manitou Springs "region accommodated twenty-five thousand tourists per season." Although that figure likely represents the overall total for months of summer bookings, we can presume that the reports of thousands of eclipse tourists—maybe well over ten thousand—is not wide of the mark.

A few days after the eclipse, the *Colorado Weekly Chieftain* reported: "Our hotel registers show the number of travelers to be large." Georgetown's *Colorado Miner* reported on August 3rd: "Tourists are here in great numbers, and all appear to be immensely pleased . . . The hotels are full and utilizing all the outside rooms they can secure now-a-days." With its regular train service and relatively close proximity to Denver, the silver-mining community was a popular site for tourists coming up to the mountains on weekends. And with its high elevation (almost nine thousand feet), it must have been absolutely packed for the eclipse.

As the 2017 eclipse approaches, I have already heard anecdotal reports from amateur astronomers that many hotels and motels in the eclipse path, from Idaho to Indiana, have been booked a year or more in advance. Based on the 1878 reports of tourists packing the limited hotels in Colorado, this is hardly surprising.

Then, like now, many of those tourists wanted to participate alongside the astronomers during the eclipse, contributing in some small way to the observations. Fortunately, thanks to a publication from the US Naval Observatory, they had an opportunity to do so.

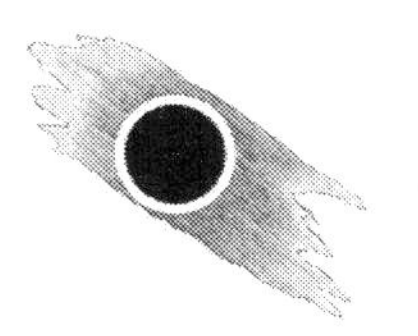

8. Among the Favored Mortals of Earth

The 1878 eclipse gave everyone, not just astronomers, a chance to make scientifically useful observations. Earlier in the year, the US Naval Observatory published and distributed a thirty-page circular titled *Instructions for Observing the Total Solar Eclipse of July 29, 1878*. This booklet was written by the Navy astronomer William Harkness (who observed the eclipse from Creston) for "persons who may witness the total solar eclipse of July 29, and who may desire to co-operate with the United States Naval Observatory, to some of the phenomena which, in the present state of science, it is most desirable should be carefully observed on that occasion."

The circular included basic instructions for observing the eclipse, such as: the importance of recording the exact time totality would begin and end in the observer's location; how to most accurately sketch the Sun's corona; and how to use a telescope to scan for the theoretical planet Vulcan during totality. It also provided blank pages in the back with sample templates on which observers could sketch their observations of the eclipse. The Naval Observatory encouraged everyone to send their drawings back to them for analysis. These drawings—done variously in pen, pencil, chalk, and oil—are still housed in the archives in Washington DC.

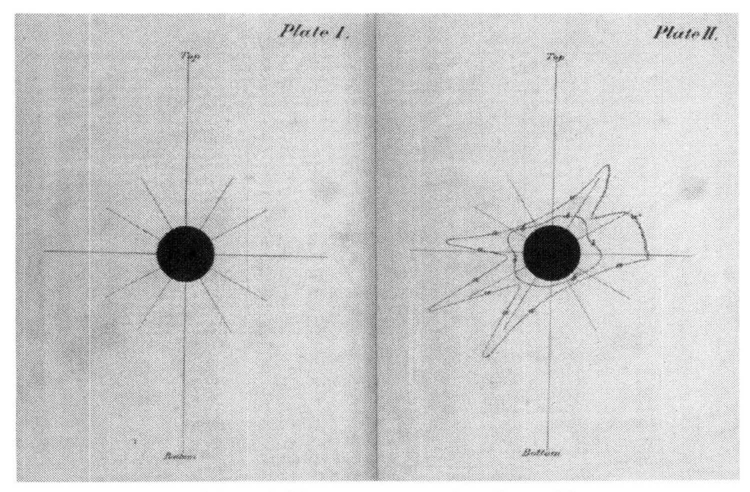

US Naval Observatory template for making sketches of the corona during totality.

This was a chance for the average person, not just the professional astronomer, to contribute to the study of the Sun. The editors of the *Georgetown Courier* became so excited that they had dreams of scientific grandeur, gushing ecstatically: "When the Navy Department issued their circular asking intelligent observers within the line of the eclipse to note the time of totality, etc., we instantly conceived the idea that at last, we were among the favored mortals of Earth, and here was the chance to strike for fame, and visions of honorary membership to the prominent scientific societies of the world."

The Naval Observatory's instructions were written to be useful to both professionals and amateurs alike, each to their ability, and with whatever equipment they had at hand. There were precedents for this type of publication. Over a hundred years prior, in preparation for the 1769 transit of Venus, the Royal Society of London distributed instructions to employees of the British East India Company scattered around the world, so even amateurs far from England could report on their observations. During a brief, and rare, astronomical event like a transit or an eclipse, the more observations made, the better, and anyone with access to a telescope, or who could draw, or even just keep accurate time, was asked to make an observation. And for those members of the public who had access to a spectroscope or other sophisticated apparatus, instructions were provided for them as well.

One way amateur observers could help the Naval Observatory with nothing more than a pocket watch, was by helping to determine the "Limits of the Shadow Path." This was the width of the moon's shadow as determined by the position of its edges as it sped across the surface of the Earth. Predicting the exact path of the moon's shadow on the surface of the Earth beforehand was not easy; in advance of the 1878 eclipse, the position predicted by British and American almanacs differed by almost four miles. The only way to know where, and how wide, the eclipse path was, required that it be measured in real time. Of course, no physical measuring stick could stretch the nearly one-hundred-and-thirty-mile width of the shadow. But it could be measured by those on the ground. All they needed were pocket watches, a knowledge of their location, and the ability to do some simple calculations.

As the Naval Observatory stated, "It is therefore important to determine accurately the true position of the path, and this may be readily accomplished by observing the duration of totality at points situated from one to ten miles within the shadow." The duration of the eclipse would appear longer to those in the exact middle of the shadow than it would to those at the very edges, because in the middle of the shadow, the Sun would be blocked by the moon for longer than it would at the shadow's edge. The shorter the time of totality appeared to an observer in the shadow, the closer to the edge of the shadow they were. Therefore, multiple observations of the duration of totality at the edges of the eclipse path would provide the data needed to later calculate the shadow's actual path.

"These observations may be most conveniently made by a party of three persons, one being furnished with a watch having a second hand, another with a common spyglass and a piece of smoked glass, and a third with a pencil and note book, to record the time."

The "holder of the watch," according to the Naval Observatory's instructions, should begin counting the seconds aloud as soon as the visible part of the Sun was "reduced to its narrowest crescent." Then, when the "last ray of the Sun has disappeared, the observer with the glass will call time, and the exact minute and second must then be immediately recorded. The observers will then wait the return of sunlight, which will seem to burst out suddenly, and the minute and second of its appearance must be noted down with the same care as the time of disappearance. The difference of the two times gives the duration of totality."

Such an operation, repeated all along the path of the eclipse by as many teams as possible, would provide the Naval Observatory with enough data points to accurately calculate the width and position of the eclipse path.

* * *

Whether professional or amateur, local or tourist, everyone had done what they could to prepare for the eclipse. Some would simply stare up in awe. Others would work diligently at their telescopes, spectroscopes, pocket watches, or sketch pads. All these preparations built anticipation for the day itself.

July 29, 1878, promised to be a momentous day, both for astronomy and for the residents of the American frontier.

As long as the weather cooperated.

PART II

OBSERVATIONS

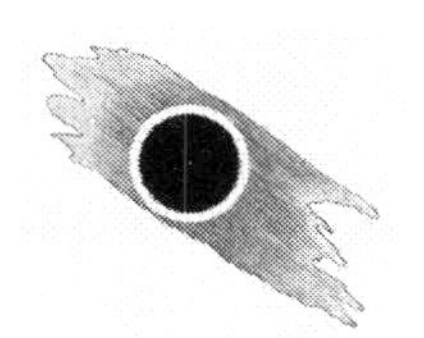

9. Weather Splendid—Clear as Hell

When the day of the eclipse arrived, all the weeks of worry about the weather were cast aside. July 29, 1878 was glorious across most of the Rockies. The morning dawned bright and clear, and the skies remained cloudless into the early afternoon. The relief felt by many was summed up by the report that came in from Walter Spencer, the *Denver Daily Times* reporter stationed in nearby Castle Rock: "Weather splendid for eclipse—clear as hell."

The *Georgetown Courier* later reported: "Monday morning, the Sun rose in a cloudless sky, and every sign bespoke a genuine Colorado summer day for the great event astronomers had promised." Yet another headline brayed, "The Eclipse! A Glorious Day for Colorado! Our State Doesn't 'Go Back' on Herself."

The day following the eclipse, the *Denver Tribune* was perhaps the most effusive. "In the presence of so many bright and shining lights of science ... THE TRIBUNE cannot let the occasion pass by without complimenting Colorado on the beautiful weather she furnished for the occasion, and the general success which attended the exhibition. It was a notable event in our history. Previous bad weather had led us to fear that yesterday might be cloudy, but Coloradans felt sure that Colorado would prove equal to the emergency and show the world that she knows how to entertain an eclipse party in good style and furnish them every opportunity for observation and scientific deduction."

With such fantastic weather, astronomers, locals, and tourists awaiting the big day all got an early start to their chosen observing sites. Although the eclipse would not begin until around 2:15 in the afternoon (the exact time of totality varied along the eclipse path), there was no time to waste. The *Georgetown Courier* later reported: "Many parties left in the morning for the mountains; Gray's Peak, Argentine Pass, Sherman, Griffith, and Leavenworth mountains, receiving full delegations."

From Wyoming to Texas, everyone had planned their day around the eclipse. In Denver, banks shut down, stores were closed, and the streets were full of awestruck observers staring upward. In Colorado Springs, a businessman taking the day off laid out a blanket, propped up an umbrella, and brought out a wooden barrel of the popular Bethesda mineral water. When a newspaper reporter wandered by and asked the man his opinion, he responded, "By Jove! Ain't it a stem-winder?"

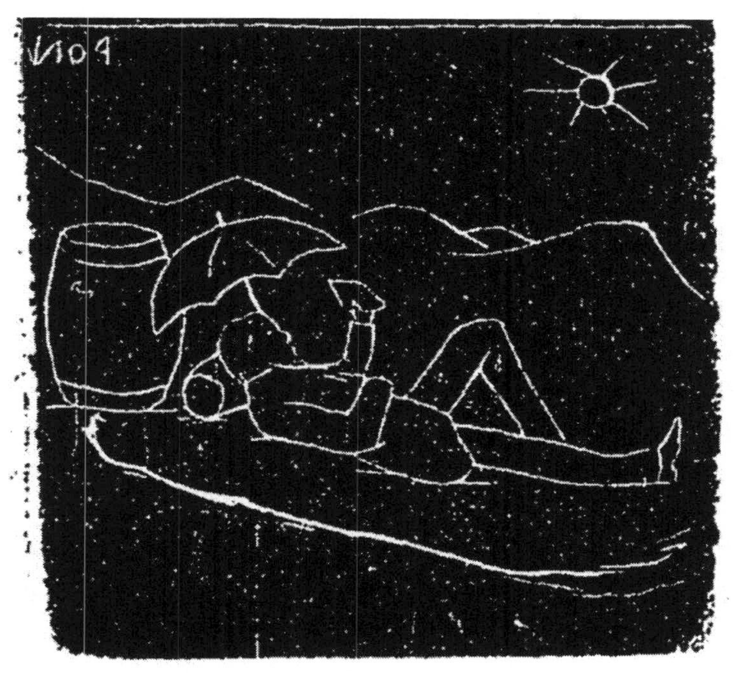

"One of our professional gentleman who viewed the eclipse from beneath an umbrella." Colorado Springs Weekly Gazette, August 3, 1878.

Many observers recorded the event in their diaries. British expatriate Frances Georgina Miller recorded how she and her husband George had just returned home in time to see the eclipse: "George & I rode over to Mr. & Mrs. McBride …extremely hot. A splendid view of the Eclipse when we got home."

An even more interesting entry comes from the diary of John L. Emerson, a miner who was excavating a promising lode he had discovered a few weeks earlier above the mountain-rimmed town of Leadville, Colorado. Emerson was a regular diarist, even though most of his entries were short, conveying the routine life of a miner (e.g. "I worked on our lode" and "anxious to get into good mineral" and "at work hard again today"). However, in his diary entry for July 29th, the monotony was broken. "Sleep in tent. Water freezes. This morning got up early and went to work getting out lumber for shaft we are going to crib it up. Worked very hard to get the timber started up in shaft. Total Eclipse of Sun."

Emerson had clearly been so impressed with the eclipse that not only did he pause his mining activities long enough to pop his head out of the ground to watch it, but he recorded the event with double underlining for emphasis, an emotional outburst not readily found elsewhere in his diary. What an exciting, memorable event the eclipse must have been for this lone miner watching from his high-altitude claim.

But despite the excitement and festival atmosphere surrounding the big day, there is one important task that anyone watching an eclipse must do: protect their eyes. In 1878, unless they were astronomers with special filters, eclipse observers had one recourse: making their own solar filters out of smoked glass.

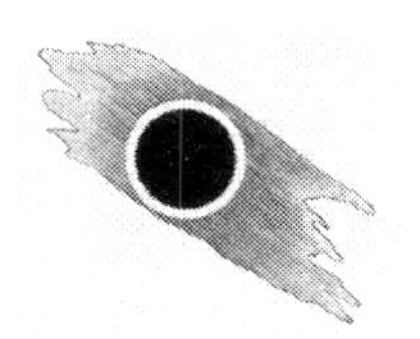

10. Candles Consumed in the Pursuit of Science

During totality, the moon blocks the Sun. But totality is so brief that, as the moon moves on, the Sun will reappear quite suddenly. Protective eyewear is a must. For more recent eclipses, observers could purchase inexpensive paper and polymer "eclipse glasses," which cost only a few dollars. But in 1878, eclipse watchers were advised to view the eclipse through smoked glass—a piece of clear glass tinted with thick smoke from a candle or oil lamp. This provided a uniform coating of soot (also called "lamp-black") through which the eclipse could be safely viewed. Smoking glass was easy to do, and on July 29th, almost everyone seemed to be doing it, although in some instances with disastrous—and dangerous—results.

Near Boulder, Colorado, a woman made a piece of smoked glass in her kitchen and then ran it out to her husband, who was working in the field. After they watched the eclipse, she returned home to find her house in flames—she'd forgotten to trim the lamp she had used for the smoking.

Less damaging (but no less startling) was Georgetown resident Dave Eldredge's attempt to create an impromptu smoked-glass filter. Caught high on Geneva Pass during the eclipse with no actual glass around, he tried using a match on the crystal of his pocket watch, "and succeeded in breaking it into a thousand pieces." Such were the dangers of using open flames. For the most part, however, smoking glass was safe and widely utilized.

Everyone, from shepherds to senators, used the stuff. James Freeman, a young sheepherder who watched the eclipse from Colorado's eastern plains, "watched the Sun carefully through a smoked glass until totality." And Senator Henry Teller, who viewed the eclipse with the crowds in the Garden of the Gods, brought with him a stack of smoked glass "neatly set in small paper frames" that had been prepared for him and his "senatorial friends" by a Denver astronomer, who used burning olive oil to create the smoke.

Other options were also employed, such as blue-painted glass that was said to be "much better than smoked glass, as it will not soil the fingers and is not expensive." A large supply of glass for smoking came from Iowa, where a hailstorm had shattered over a thousand greenhouse glass panes. Making lemonade from lemons, that glass was shipped to Colorado to be cut, smoked, and sold.

All along the eclipse path, noted the *Rocky Mountain News*, glass was at a premium: "While the professionals, with their sails trimmed, calmly awaited Luna's approach, average citizens were frantically engaged in hunting pieces of broken glass in back yards and burning it and their fingers over dubious light on the kitchen table. The street vendors' stock of the dusky article was soon exhausted, and the demand continued up to the first moment of contact."

Lake City's *Silver World* wrote, "It is not too much to say that every man, woman, and child in town had a

Citizens from all walks of life viewing an eclipse through pieces of smoked glass. (Harper's Weekly, November 4, 1865.)

piece of smoked glass or something of that nature, through which the eclipse was taken in." It was the same in Pueblo, Colorado. "During the entire morning, about one half of the population seemed to be engaged in the interesting and scientific operation of smoking glass. Broken glass was at a premium, and numerous candles were consumed in the pursuit of science."

Observers also got creative, embedding their smoked glass into pieces of cardboard or boxes, which they put over their heads. Another "enthusiastic scientist" put two green-glass wine bottles to his eyes, while another prankster used the soot from the smoked glass on a device that "blackened the nose of anybody who looked through it. On the streets, black noses and faces daubed with perspiration and soot were the rule."

As far as safety was concerned, the *Saguache Chronicle* reminded readers that "glass should be smoked beforehand," and that each observer should have his own piece. Parental advice was also offered. "Let mothers see that each of the little ones are provided with a piece of glass; and be sure that you do not suffer them to look at the Sun with the naked eye; by so doing, permanent injury to the eyes may be prevented." As an alternative to smoked glass, the same article suggested using a washtub filled with clear water, to view the eclipse in reflection.

Two days after the eclipse, Denver's *Daily Tribune* quipped that "frugal individuals are saving their smoked glass for the next total eclipse." And on August 1st, Pueblo's *Colorado Weekly Chieftain* got entrepreneurial. "Several acres of smoked glass, which will be valuable to observe the next total solar eclipse, for sale on reasonable terms at the *Chieftain* office."

Surprisingly, some nineteenth-century astronomers did occasionally coat their telescope's objective lens (the large, outer lens in most refracting telescopes) with lamp-black. But most of them used more sophisticated filters, like a "Herschel wedge," a prism designed to reduce the intensity of the Sun's light by up to 95 percent. Such a wedge was used in the Naval Observatory telescopes provided to the eclipse expeditions. The professionals, like the amateurs, had to be careful not to blind themselves.

So, as the locals and tourists smoked their glass in a spirit of festive anticipation, across the Rockies, the mood of the visiting astronomers was much more somber. For everyone else, the eclipse couldn't come fast enough. But for the astronomers, every second before totality was precious. No time was wasted as they hurried to check and recheck their instruments, making sure all was ready.

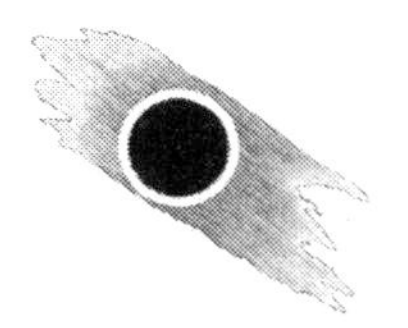

11. INTENSELY BLACK

Along the eastern peaks of the Rocky Mountains and out on the Great Plains, the scattered line of astronomers readied their telescopes, spectroscopes, and cameras. They were giddy with anticipation, and also a little on edge. The diligent ones had spent days, even weeks, readying their equipment. George Barker, in Wyoming, recalled, "The time from the date of our arrival at Rawlins until the eclipse was occupied in setting up the instruments, in getting them into adjustment, and in practice with them." To understand the rough conditions in which the astronomers were working, consider William Harkness's description of his team's campsite at Creston.

"Our camp consisted of the postal car in which the instruments were brought from Washington, which served as office and dormitory; a temporary observatory, built of rough boards; a mess tent; a cooking tent; and a small A-tent for the two soldiers. The temporary observatory was erected before our arrival, the necessary materials and workmen having been brought from Rawlins. It was a building measuring 12 by 22 feet, 7 feet high at the eaves, and 10 feet high at the ridgepole. The eastern side of its roof was of boards, while the western side was of canvas, so arranged as to be readily removed when the instruments were in use. All these instruments were supported on blocks which passed through the holes in the floor, and rested firmly upon the ground, thus securing the greatest steadiness."

Along the path of totality, from Wyoming to Texas, the astronomers' stations were equally makeshift, varying only in matters of degree. There were tents on hillsides on the outskirts of Denver, walls of lumber built behind sandy dunes in Separation, and restaurant chairs set next to brick chimneys on hotel roofs in Central City. Under such conditions, teamwork was essential. Each member of an observation team had an assigned role.

In the afternoon of July 29th, when the eclipse finally started, an assistant in each of those camps was to call out, "First contact!" and record the exact time. The eclipse shadow started in Siberia and ended in the Atlantic Ocean. The astronomers in Wyoming were the first to see "first contact"—the moment when the moon initially crosses the disk of the Sun. But it was another hour and a quarter until second contact, giving the astronomers time to fidget with their instruments beneath the darkening sky. They watched in nervous anticipation as, inch by slow inch, the moon blanketed the Sun.

Just before the Sun was completely hidden, they saw what looked like an arc of glowing pearls along the moon's outer edge. "Bailey's beads," they're called: sunlight broken into fragments as it passes between gaps in the moon's crater-hewn ridges.

Then the Sun was gone.

"Second contact" was announced. Day turned to night, and the stars and planets suddenly, majestically, appeared. No time to gawk, however—the astronomers, many of who had spent the previous few minutes in a dark tent to adjust their eyes to the onset of night, worked hurriedly to accomplish whatever it was they'd come to do. They had three minutes at most, so time was of the essence.

Eclipse camps maintained as much silence as possible. They'd hear the voice of the designated timekeeper, whose sole task was to count the passing seconds. At the Rawlins camp, the timekeeper was Henry Draper's wife, Anna, whose "clear and distinct counting" rose above the hurried, hushed conversations between astronomers double-checking each other's observations. Anna, her hand grasping her timepiece, counted off one hundred and sixty-five seconds. Then, all too soon, the first crack of sunlight reappeared.

"Third contact!"

And with that, totality was over. They observed the moon's shadow rushing madly away over the ground, and the normal colors of daytime slowly returning. The Sun grew brighter for another hour or so as the moon moved on. Finally, as it left the face of the Sun completely, "fourth contact" was announced, and the eclipse was officially over.

During the brief period of totality was when the important work was done. The astronomers kept their eyes pressed to their instruments, which meant missing the spectacle of the writhing, luminescent corona—the most spectacular part of the eclipse. They couldn't afford to simply watch the eclipse in its entirety; science demanded they each concentrate on a small piece of the whole. "It required not a little self denial to ignore all the grand scenic effects of the eclipse," wrote Lewis Swift, observing from Denver.

Sketch of the corona during totality from the summit of Pikes Peak. Samuel Langley, New Astronomy (1888).

But some of them just couldn't resist. The physicist George Barker stole "a few seconds of the precious time to observe the eclipse with the naked eye." He was stunned. "The moon appeared intensely black, surrounded by a pinkish halo. The amount of light emitted by the corona was a surprise to me, equal to that given by the moon when ten days old."

The corona—Latin for *crown* or *wreath*—is normally hidden from sight by the Sun's overpowering radiance. But during totality, the moon blocks that light, revealing the corona in all its shimmering glory—a pulsing, pearly whiteness, often infused with subtle shades of blue, green, and red.

One of the devices used during the eclipse to investigate the corona was a spectroscope. Spectroscopes analyze sunlight, using prisms to separate it into its component colors. With spectroscopes, astronomers in 1878 were investigating the light of the corona for patterns of black lines along the visible (and invisible) spectrum, which were markers for the chemical elements of the Sun's gaseous outer layer.

When astronomers look at incandescent light with a spectroscope, they see patterns of lines: bright lines from glowing gasses, or dark lines (also known as Fraunhofer absorption lines) when light from glowing solids is passed through a cooler gas. Every chemical element—hydrogen and helium, for example—has a distinct pattern of both types of lines, which had first been determined in 1859 by the German physicist Gustav Robert Kirchoff. But in 1878, the exact composition of the corona, or whether it was even a hot gas at all, was still a mystery (although during the 1868 eclipse over India, Norman Lockyer and Jules Janssen used a spectroscope to discover helium in the Sun, before it had been discovered on Earth). This helps explain why Henry Draper's primary objective when observing the 1878 eclipse from Rawlins was to use spectroscopes and cameras to learn "whether the corona was an incandescent gas shining by its own light, or whether it shone by reflected sunlight." Using a spectroscope on the Sun was like doing a chemistry experiment from ninety-three million miles away.

Photographing the corona was another objective of the expeditions. Photography had advanced considerably by 1878, and it was hoped that astronomers would be able to get static images of the corona that could be studied later—capturing totality permanently, so to speak. Specially designed cameras were attached to telescopes and spectroscopes, but they were slow, requiring long exposures on wet collodion plates. In the few minutes of totality, even a skilled photographer had time to make only a handful of images.

Photos of the corona were made in Rawlins and La Junta. And a local portrait photographer in Central City also made an image with his commercial camera, which many judged to be better than many of the photos made by the astronomers.

For these astronomers, the eclipse, specifically the solar corona, was a thing to be analyzed, broken down into its constituent parts. They split coronal light with spectroscopes, measured its heat with thermopiles, and captured its essence on photographic plates. But for the non-scientists, the eclipse was observed as a unified phenomenon. They enjoyed it in its entirety instead of in small, measurable pieces. To one of those holistic observations I now turn.

12. A Sublime and Joyous Vision

When watching an eclipse while standing on flat ground, you do not get a sense of the incredible size of the moon's shadow as it races across the surface of the Earth. You are simply surrounded by it, engulfed in darkness with no sense of the outer limits of the sudden night. But from high above the surface, you would have a different perspective.

The top of Pikes Peak rises eight thousand feet above the Great Plains, and over fourteen thousand feet above sea level. From that height, and on a reasonably clear day, you can see all the way to Kansas. I know—I used to work as a mountain bike guide on

Pikes Peak during my summer vacations. We would be up on the summit by seven a.m., well before most of the other tourists. Even on the cold days, I always took a moment to look out at the view—the same view that inspired Katherine Lee Bates to write the song "America the Beautiful."

That view was always stunning. For me, the most beautiful days were when there was a layer of clouds far below, like a cotton blanket floating at ten thousand feet, stretched over the bed of Earth. On those days, standing high above the endless clouds, you felt as though you were on an island in the middle of an ocean. But on clear days—which were the norm—when there was not a cloud in sight, as I looked down upon the plains to the east, and out toward the Rockies to the west, I felt like I was on top of the world. Whatever was happening down below seemed distant, small, far away. It was wonderfully peaceful.

I suspect that the one exception to that feeling would be the sudden appearance of the shadow of the moon, which would dwarf even the massive grandeur of that high peak known as "America's Mountain."

From the summit of Pikes Peak, the extent of the lunar shadow on the surface of the Earth below would be much easier to see in its entirety. In 1878, one lucky resident from Manitou Springs, a small town tucked into a narrow valley at the base of Pikes Peak, was invited to join the official observing party on the summit. Not tasked with any scientific job, this unnamed person, identified only as a local "gentleman," was allowed to enjoy the view and take in the spectacle. Fortunately for us, he wrote down what he saw. His account is beautifully written, the sort of thing an historian loves to come across because it is so evocative of a particular moment in the past.

This person's account is worth quoting at length, because more than any other record of the eclipse, it gave readers—then, and now—a powerful, panoramic sense of what the eclipse looked like from high above those frontier cities, towns, ranches, and farms.

2

THE ECLIPSE.

A Grand Spectacle.

Colorado College Observation.

A View From Pike's Peak.

Eclipse Notes.

Sketches by our Special Artist.

The Eclipse from Pike's Peak.

The view of the eclipse of Monday from the summit of Pike's Peak—over 14,000 feet above the level of the sea, and probably the highest point from which a total eclipse has ever been observed—is said, by those who were so fortunate as to see it, to have been very fine and in some respects very remarkable. A gentleman who was one of the party from Manitou invited as the guests of Gen. Albert J. Myer, Chief Signal Officer of the army, to witness the eclipse from the signal station on the summit, has given us the following description of what he saw:—

Eclipse reporting. Colorado Springs Weekly Gazette, August 3, 1878.

When viewed from the top of Pikes Peak, wrote our anonymous gentleman, "the most remarkable phenomenon was the approach of the shadow of totality from the north, its swift passage, and the sudden burst of sunlight which followed.

It must be remembered that the movement of the shadow from north to south was at the rate of about 30 miles per second, and it was necessary that one should have stretched beneath him a wide expanse of the Earth's surface, in order that the eye could seize with a glance the line of the approaching shadow, and follow it as it rushed away to the southward. This advantage an observer who stood upon [Pikes] Peak possessed.

Without the aid of a [telescope] the eye could distinctly note the shimmering of the bright sunlight on the mountains more than 100 miles to the northward. Suddenly we observed them disappear and a great wall of darkness, stretching out on either side as far as the eye could reach, concealed them from us. With inconceivable rapidity the shadow swept towards us, its front a clear black line bordered with a fringe of yellow. It hid from sight range after range of the more distant mountains and quickly covered with a ghastly pall the peaks and foot-hills and plains beneath us."

I'll interrupt our gentleman's account at this point to note that the "shimmering" of sunlight and the fringe of yellow he described were probably what are known as "shadow bands." These are narrow, undulating bands of dark and light that arrive just before, and follow immediately after, the moon's shadow during a total eclipse. And now that shadow is approaching, so let us return to our gentleman on the mountain.

"When the shadow reached and enveloped us, the eye could dimly outline the nearer mountains, and could single out with strange distinctness the houses and farms in the valley below. The sky overhead seemed heavy and leaden, and every visible object was pallid and ghastly. The very shadow seemed tangible, but the horizon all around us was brightly illuminated by flashing rays of red and yellow lights like those of the Aurora. While we on the Peak were still enveloped in the depth of the shadow its upper line passed over the far-off range to the northward, and the clear sunlight struck the mountains, and away beyond and through the darkness they burst suddenly into view. In an instant other and nearer mountains appeared and then the dimly shaded parks and the wooded Divide were bathed in sunlight, and the shadow rushed past us."

Our mountaintop tourist concludes his description with a reference to the concepts of the *sublime* and the *beautiful*. In the nineteenth centuries, picturesque images, whether drawn with eloquent words or a painter's brush, contrasted the *sublime*—that which can inspire awe, or even terror—with the *beautiful*—that which brings overwhelming joy. See here how our gentleman finishes his description of the eclipse with these aesthetic sensibilities.

"During the period of darkness the view around us was weird and terrible, but the sudden burst of sunlight which appeared upon the distant mountains as they seemed to spring up instantaneously from the bosom of the Earth, was one of the most sublime and joyous visions that it is ever given to mortal eyes to witness. It brought a sense of relief and delight, and no one who saw it can forget it."

Fortunate indeed was our anonymous gentleman, taking in the grandeur of such a magnificent sight while the professional astronomers around him had their heads down at their instruments cold brass eyepieces, or sketched madly at their pads of paper while they squinted up at the eclipsed Sun.

* * *

Speaking of those astronomers, there were dozens of eclipse expeditions spread along the shadow path, each doing their very best to snatch something useful from the Sun in the few unforgiving minutes of totality. Some of those expeditions were funded by Congress. Others were supported by universities, or were privately funded. A few came from Britain, including a group from the Royal Astronomical Society. But it would be tedious if I tried to recount them all.

So in the next four chapters I will describe a few of the most remarkable expeditions that came to the Rockies in 1878. Each had a unique approach to observing the eclipse. One took place at an extreme altitude, another combined celebrity with genius, a third was radically progressive, and the forth was hunting for a ghost around the Sun.

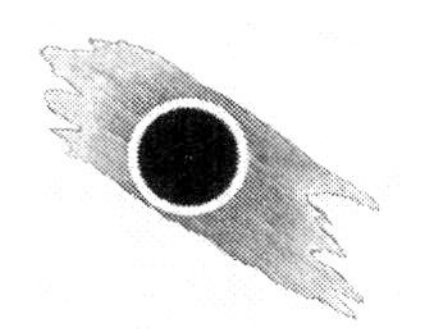

13. Difficulties of a High Mountain Peak

While the rest of the frontier was worried about the eclipse being ruined by bad weather, high on top of 14,115-foot Pikes Peak, expedition leader Samuel Pierpont Langley was experiencing it first hand.

A little more than a week before the eclipse, Langley and his team scaled Pikes Peak and set up their observatory, which included the large telescope, on loan from the US Naval Observatory, that had arrived in nearby Colorado Springs packed into eleven boxes. Those boxes (weighing over half a ton), along with a few other instruments, reportedly took a team of twelve donkeys a total of five trips to bring to the barren, rocky summit, while Langley and his team waited anxiously on the side of the mountain for a few days as their equipment was delivered piecemeal. "The path was so narrow," recalled Langley, "with such sharp turns among the rocks and such opportunities for accident, that it seemed doubtful whether the boxes, when strung between two animals walking in single file, could be carried safely."

After a few days (each round trip from Colorado Springs to the top of Pikes Peak was roughly thirty miles, with a climb of nearly eight thousand feet), all the equipment was on the summit, and they began making preparations for the eclipse. Although the weather above fourteen thousand feet is always extreme, it was the middle of July, and if there was ever a chance for mild conditions, it was then. Yet the weather in the days leading up to the eclipse was particularly nasty.

Langley, a hearty man who was also an aviation pioneer, fought constantly against the horrendous winds that battered his tent, which "roared with a noise like that of a loose sail in a gale at sea." One morning, just a few days before the eclipse, he woke to find a ten-inch drift of snow lying next to him on his pillow. This was *inside* his tent. But the choice to observe from Pikes Peak had been his, and he would see it through. He later explained his rationale for "choosing to encounter the difficulties of a high mountain peak." He wanted to investigate "the corona through the peculiarly dry air of Colorado, and from an elevation of 14,000 feet."

The top of Pikes Peak was at the time controlled by the US Army, and its Signal Service maintained a rough stone meteorological station at the summit, with a telegraph line connecting it to Colorado Springs some eight thousand feet below. Upon arriving on the summit, Langley's party was disappointed to find that there was no accommodation for them inside the signal station as they had been promised. The officers in the station had received no advance notice of Langley's arrival. Even if they had, there was no room in the small building for the observers and their instruments.

Langley's team was therefore forced to set up tents, and do their best to protect their equipment from the harsh conditions. This involved rather extreme and creative efforts, such as when Langley used a can of lard from the signal station kitchen to coat the telescope's steel parts, to protect them from mountaintop moisture. And Langley's brother, John, had to construct a forty-foot-long tube for his photometric apparatus, stabilizing it with piers built from the rocks and sticks he scrounged from Pikes Peak's summit. The astronomers slept out on the summit, too, in their canvas tents that barely qualified as shelter, incapable as they were of keeping out the elements.

But even if they'd had decent shelter, observing a solar eclipse from fourteen thousand feet above sea level is no easy task. They faced two great difficulties: the weather and altitude sickness. A perfect example of the unreliable weather common on Pikes Peak is found in Langley's description of his efforts, on July 24th, to set up his telescope (known as an "equatorial" telescope, for the way in which it is mounted).

Langley drew this sketch of the phenomenon known as "Brocken specters": himself, his brother John, and Cleveland Abbe on top of Pikes Peak, when their shadows were cast on nearby clouds by the setting sun.

"The day was passed in fruitless attempts to adjust the equatorial. In the morning, the canvas which covered it was frozen and loaded with hail. A little later, the Sun shone out suddenly and with surprising warmth, turning the hail to water. I commenced unwrapping the canvas, and was lifting it off, when the Sun disappeared as suddenly as it came out, and before

I could put the cover on again, it was hailing once more, and we were involved in dense cloud. The cloud was continuous, except for several brief moments of sunshine, during which I uncovered the instrument several times to no purpose."

Of altitude sickness, Langley reported: "It is one thing to stay [on the summit] for an hour or two, and another to take up one's abode there and get acclimated. Mountain-sickness is something like violent sea-sickness, [along] with the sensation a mouse may be supposed to have under the bell of an air pump."

That combination of nausea and suffocation was particularly tough on Langley's companion, Cleveland Abbe. Abbe, a bookish man who was probably more at home in cities than on mountaintops, was a meteorologist and early advocate of using telegraphs to transmit weather reports. The day before the eclipse, Abbe had to be carried off the summit. He ended up observing the eclipse at an elevation of ten thousand feet. "Professor Abbe's condition gave us cause for alarm. . . .On the evening of the 28th, [he was] put into a litter and carried down to a lower altitude, where his recovery was rapid."

Thus, for a party of astronomers camped high atop Pikes Peak, the weather during the week leading up to the eclipse (to say nothing of the lack of oxygen) did not inspire confidence. "Hail, rain, sleet, snow, fog, and every form of bad weather continued for a week on the summit," recalled Langley. But on the day of the eclipse, the morning dawned clear and cloudless, to his tremendous relief.

Langley's telescope was a heavy brass instrument with a five-inch aperture (the width of its largest, or "objective," lens) and about a nine-foot focal length. He also brought a spectroscope, a borrowed heliostat (a mirrored device for redirecting sunlight), and a photometer (to measure the intensity of sunlight). Unfortunately, the telescope from Washington arrived at Langley's Pennsylvania observatory mere hours before he had to turn around and put it on a train to Colorado. He had hoped to attach the spectroscope and heliostat to it, but he ran out of time.

To make things worse, upon arriving in Colorado Springs, Langley realized that the Naval Observatory had neglected to include important parts of the heliostat they loaned to him. So a week before the eclipse, Langley was about to climb to fourteen thousand feet without all the necessary equipment to execute his intended observations. He resolved to make the best of it, "with the hope that something might yet be done."

Because Langley's planned observations were thwarted by incomplete equipment, and made difficult by extreme weather, the scientific results of his expedition were not at all what he had hoped. He could not analyze the Sun's corona with the spectroscope and heliostat as planned, so he focused instead on making an accurate drawing of the corona.

Having seen solar eclipses elsewhere in the world, he was immediately struck by how far the coronal streamers—the gauzy rays of particles streaming through the Sun's corona—extended outward from the Sun. At previous eclipses, the streamers had been significantly shorter, but in 1878, he measured them at twelve times the Sun's diameter—more than ten million miles! They were so extensive that they were visible for many minutes after totality ended. It was extraordinary, and Langley's drawing of the streamers is still the defining image from that eclipse. Most of the other astronomers agreed that the streamers were more extensive than at any other eclipse in living memory. This is where Langley's expedition turned out to be a success, because he realized that some phenomena were best observed at high altitude. His sketch was arguably the most impressive rendering of the streamers out of all the images made in 1878, whether photographs or other observers' drawings.

But after a week of fighting fickle weather and altitude sickness, Langley concluded that fourteen thousand feet was just a bit too high for effective astronomical work. Observing from slightly lower elevations, somewhere around ten or eleven thousand feet, would be better. The air would be easier to breathe, and trees nearby would help provide much-needed shelter from the wind.

Despite the difficulties of high-altitude observation, Langley and his team considered their expedition a success. And yet, there had been one other disappointment. While on Pikes Peak, Langley had hoped to test a recent invention by Thomas Edison: the tasimeter. But Langley never received it. At the last minute, Edison decided to go west and see the eclipse himself, and he brought his tasimeter with him.

14. The Celebrated Inventor

More than any other participant in the 1878 eclipse, Thomas Edison's appearance was the most covered by the press. Although he was not an astronomer, he brought with him a new device he'd created in his Menlo Park laboratories. The "tasimeter" was a highly sensitive infrared detector, which Edison claimed could measure heat to a millionth of a degree. Astronomers had long wanted something sensitive enough to measure the heat of the corona accurately.

This invention had come about the previous year, when Samuel Langley sent Edison a device for measuring infrared radiation called a thermopile, hoping the great inventor could improve upon its sensitivity. Langley wanted to confirm that the corona visible during a total solar eclipse was in fact a feature of the Sun, and not the illumination of some unknown atmosphere of the moon, backlit by the Sun. Heat was one way to make this determination. If Edison "could make something ... say one hundred times as sensitive," Langley wrote, "and capable of being uninfluenced by alien radiations, you would not perhaps produce anything commercially paying, but you certainly would confer a precious gift on science."

Edison took up Langley's challenge and responded with the tasimeter. The name came from the Greek words for *extension* and *measure*, as the device was designed to "measure extension of any kind." Through various iterations, he claimed to achieve infinitesimal sensitivity: first, 1/24,000 of a degree Fahrenheit, then 1/50,000°F, and then 1/500,000°F. Finally, in July, just before he left for the Rockies, he made the astounding claim that he had achieved a sensitivity with the tasimeter of 1/1,000,000°F. "It is so sensitive," claimed the *Laramie Daily Sentinel*, "that let a person come into the room with a lighted cigar, and it will drive the little animal wild."

In June, as Edison was hyping the tasimeter's sensitivity, Langley asked him to send one to him at the Allegheny Observatory in Pittsburgh. A few weeks later, he had to ask again. By July 4th it had still not arrived, and the eclipse was less than a month away. "Send by express to Allegheny," Langley hastily telegrammed. "I leave Monday."

He never received it—it is more accurate to say that Edison probably never sent it—and so Langley made his own measurements of solar radiation from Pikes Peak with one of his less sensitive thermopiles. Edison did apparently send a tasimeter to the astronomer Charles Young, who took it with him to Denver.

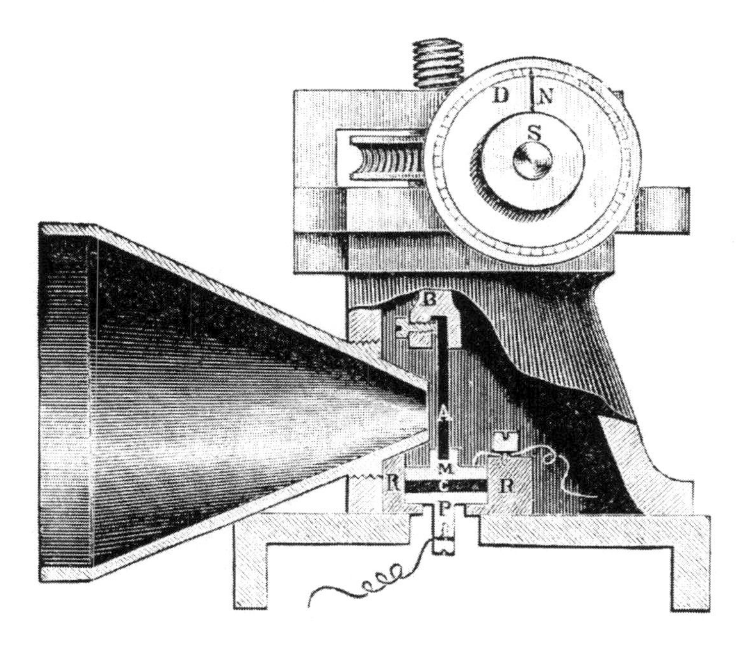

Thomas Edison's tasimeter (cross section).

It was George Barker who invited Edison to join the Draper expedition in Wyoming. Edison was apparently in need of a vacation, the recent work at his laboratory at Menlo Park having left him mentally exhausted. When he left for Wyoming, the press was there to cover his departure. Ever the showman, he perhaps wanted to be the one to first publicly demonstrate the tasimeter during the eclipse. The next day, the tasimeter got front-page coverage in New

York. The Union Pacific Railroad, thrilled to be transporting "the celebrated inventor and telegrapher," gave Edison free passage and unlimited use of their telegraph lines to report his findings.

His celebrity status followed him all the way to Rawlins, a railroad town of just eight hundred inhabitants. He arrived in time to get one of the last rooms in the hotel. That night, a drunk man ("tall, handsome, and dressed in Western style") pounded on Edison's door. The man introduced himself as "Texas Jack." Mr. "Jack" wanted to meet the inventor he'd heard so much about, and reciprocated by trying to impress Edison with his cowboy marksmanship. He flung open Edison's window and fired at a nearby weather vane with his six-shooter, until Edison was able to convince him to let them all finally get some sleep.

At the time, Edison was only thirty-one ("young looking," wrote one Chicago reporter, "with a smooth-shaven face, a straw hat, and a linen duster"), and was at the height of his inventive powers. He had been talking about his tasimeter for months, to the point where he may have overhyped it.

Edison was not an astronomer, but at Rawlins, he worked alongside some of the best. Henry Draper (an astrophotography pioneer) and Norman Lockyer were there. Lockyer, a leading proponent of the "new astronomy," was keenly interested in the tasimeter. A few nights before the eclipse, they all tested the device on the star Arcturus. Lockyer was impressed. "It is truly a very wonderful instrument, and from the observations made last night on the heat of Arcturus, it is quite possible that Edison succeeded in his expectations."

The day of the eclipse—like previous days—saw the persistence of the steady winds for which Wyoming is famous. Because Edison had arrived late to Rawlins, the sheltered observing sites had already been claimed. He was forced to set up in a chicken coop, positioning his tasimeter behind the telescope he would aim at the corona during totality. But the incessant wind, and the dust it kicked up, made it very difficult for him. He was a brilliant inventor, but no astronomer, and he lacked the experience his companions had obtained on previous expeditions.

Lockyer, Draper, and the others had been busy at their own instruments all day in order to be ready for the eclipse, but when totality began, Edison had *still* not fully adjusted his hyper-sensitive tasimeter. The few minutes of totality sped by, and because of the immediate darkness, the chickens, whose home Edison had invaded, suddenly came home to roost!

Or at least that's the story the press told. There is considerable debate among historians about the truth of eclipse-crazed chickens disrupting Edison's experiment. What is not debated, however, is that Edison failed to get accurate readings of the solar corona with his tasimeter. He was not ready in time, and despite his last-ditch effort to aim his tasimeter-telescope at the "edge of the luminous corona," he was still unable to get data useful to astronomers. He *did* get data, in so far as the tasimeter briefly detected coronal heat, but ultimately declared it unusable. "My apparatus," he later wrote, "was entirely too sensitive and I got no results." Besides, the heat of the corona had already been detected by the Italian Luigi Magrini, years before Edison was even born.

Charles Young, who brought one of Edison's prototype tasimeters with him to Denver, had to set it aside and use a standard thermopile. He had the same problem as Edison: the tasimeter was simply too sensitive to be useful during the eclipse—it was "erratic and difficult," in the words of historian and astronomer John Eddy. "Its response time was very slow. It gave non-repeatable results. It was entirely too delicate to be of practical use."

What Langley had wanted from Edison was a device that could actually *measure* heat, not just detect it. So, a few years later, he invented what Edison did not—the "bolometer," which is still used to measure radiant energy, like heat. Meanwhile, Edison gave away the rights to manufacture the tasimeter for the benefits of science, and moved on to other projects, like the electric light bulb, which he began working on the month after the July eclipse.

Edison's trip out west wasn't a total waste. After some time in California, he returned to Rawlins to hunt and fish. And his celebrity quickly overshadowed his tasimeter results, anyway. The press eagerly reported on the number of fish he caught, and the elk, deer, and antelope he bagged. Wyoming couldn't have been happier to have hosted the renowned inventor.

Not so Colorado. Perhaps the biggest disappointment for that state in the months leading up to the eclipse was that Edison chose to observe from Wyoming. In an article bearing the headline "Edison's

Go-By," the *Denver Daily Tribune* reported: "It has been confirmed that Professor Edison will not visit Colorado for the purpose of observing the coming solar eclipse. The *Tribune* regrets this as much as anyone can, because we were anxious that Denver should enjoy the distinction that such a visit would have given it."

Denver did host another famous astronomer, however, although her fame was for reasons we would consider progressive, rather than inventive.

15. All Good Women, and True

Maria Mitchell was a true pioneer. In 1847, she became only the third woman in history to discover a comet—a rare enough feat for any astronomer, male or female. But she was also the very first woman elected to membership in the American Academy of Arts and Sciences, and was the first female professor of astronomy in the United States.

At the time of the 1878 eclipse, Mitchell was a professor at Vassar College in New York. Vassar was still an all-women's school, and it was the first such degree-granting college in America. Mitchell was a strong advocate for women's education, so when she decided to go to Colorado to view the eclipse, it was no surprise that her party of fellow observers consisted entirely of women, all of who were her former students.

"We started from Boston as a party of two," she recalled. "At Cincinnati, a third joined us. At Kansas City, we came upon a fourth who was ready to fall into our ranks. And at Denver, two more awaited us; so we were a party of six—'All good women and true.'"

Due to a "railroad war" between rival companies, Mitchell and her girls lost their telescopes and other equipment, their baggage being retained in Pueblo, Colorado, while they went on to Denver. They arrived in Denver only six days before the eclipse, with almost none of the equipment they needed to observe it. They desperately combed the city for any telescope they could rent or buy, but with no luck. Denver, flush with new wealth from silver discovered in the nearby mountains, was bustling. Locals had already snatched up any telescopes to be had. So Mitchell fired off a flurry of telegrams, and her girls waited on the station platforms as each new train pulled in. Finally, after two frantic days, their equipment arrived. It was Friday, July 26. The eclipse was Monday.

And then it started to rain.

"It rained hard on Friday," Mitchell lamented. "Nothing could be done. It rained harder on Saturday. It rained hardest of all on Sunday, and hail mingled with the rain. But Monday morning was clear and bright." With the eclipse starting around two p.m., they had relatively little time in which to get ready. Setting up their tent on a hillside, they turned their three telescopes skyward and divided their responsibilities. One of the women kept time, another made naked-eye observations, and a third was ready with pencil and paper to make sketches. Mitchell and the remaining two were at the telescopes.

Maria Mitchell and her observing team awaiting totality in Denver. Courtesy Archives and Special Collections, Vassar College Library, #08.09.04.

When the eclipse began, Mitchell wrote, the moon became "densely black . . . a black notch on the burning disc." While they awaited second contact, a photographer took their picture. Then, in hushed silence, they awaited totality. Mitchell captured the moment perfectly. "As the last rays of sunlight disappeared, the corona burst out . . . so intensely bright that the eye could scarcely bear it; extending less dazzlingly bright around the Sun for the space of about half the Sun's diameter, and in some directions sending off streamers for millions of miles."

As totality ended, Mitchell and her observers made notes in silence, until someone nearby shouted, "The shadow! The shadow!" Mitchell and her girls looked up from their notebooks to see the eclipse shadow gliding swiftly away, "The dignified march of the moon." Looking around—the high Rockies to the west, the endless rolling plains to the east—the vista was otherworldly. Denver and everything around them looked surreal in the altered light of the reemerging Sun. "What a strange orange light there was in the north-east! What a spectral hue to the whole landscape! Was it really the same old Earth, and not another planet?"

Mitchell concluded her recollection of the 1878 eclipse with her belief that women were as good—if not better, in many respects—than men at astronomical observation. Using analogies that we would find quaint today (if not somewhat regressive), she praised the nineteenth-century upbringing of a girl, including embroidery, as skills that are easily transferable to the science of astronomy.

"Nothing comes out more clearly in astronomical observations than the immense activity of the universe.

Observations of this kind are peculiarly adapted to women. Indeed, all astronomical observing seems to be so fitted. The training of a girl fits her delicate work. The touch of her fingers upon the screws of an astronomical instrument might become wonderfully accurate in results; a woman's eyes are trained to nicety of color. The eye that directs a needle in the delicate meshes of embroidery will equally well bisect a star with the spider web of the micrometer. Routine observations, too, dull as they are, are less dull than the endless repetition of the same pattern in crochet work."

Mitchell believed a girl's domestic training transferred directly to astronomy. "A girl's eye is trained from early childhood to be keen. The first stitches of sewing-work of a little child are about as good as those of the mature man. The taking of small stitches, involving minute and equable measurements of space, is part of every girl's training; she becomes skilled, before she is aware of it, in one of the nicest peculiarities of astronomical observation."

Despite their skill, Mitchell and her students failed to locate a planet known as Vulcan, which was believed by many to circle the Sun inside the orbit of Mercury. Mitchell herself seemed skeptical of Vulcan's existence; she certainly did not see anything like an intra-Mercurial planet during the few minutes of totality. But a hundred and fifty miles to the northwest, two astronomers, having just witnessed the same eclipse at their remote camp near Separation, were convinced they had just discovered a new planet.

16. The Ghost Around the Sun

The planet Vulcan does not exist (except, I suppose, in the Star Trek universe). In the nineteenth century, however, many astronomers were convinced that a planet called Vulcan did exist—not in science fiction, but inside the orbit of Mercury, so close to the Sun that it was hidden from us by the solar glare.

The only chance anyone would have to spot Vulcan was during a solar eclipse. Then, and only then, would the Sun be obscured enough to see Vulcan, which would appear as a small point of light, like a star. And even *then*, the planet would have to be off to one side of the Sun during the eclipse, instead of in front or behind where it would remain concealed. It was like trying to find a needle in a haystack, with the added uncertainty that you weren't even sure if the needle was actually there in the first place.

Where did the idea of Vulcan come from, and why was there so much uncertainty about its existence? To answer that, a little bit of planetary history, as well as a double dose of gravity *and* relativity, is in order.

The planet Mercury, the innermost planet in our solar system, has a slight "wobble" in its orbit. The astronomical term is "precession of the perihelion," but in simple terms, it means that something appears to be tugging on it, gravitationally speaking, while it orbits the Sun.

All planets orbit the Sun in an *ellipse*, a particular kind of oval. The point in their orbit at which they come closest to the Sun is called their perihelion. In general, perihelion is a fixed point in space; that is, a planet would be expected to have perihelion at the same point each time it came around the Sun. But Mercury does not. The point of its perihelion relative to the Sun shifts slightly with each trip. It was assumed that the gravity from another planet—an *unknown* planet between Mercury and the Sun—was the cause.

There was a precedent for this assumption. In 1846, the planet Neptune was discovered because Uranus exhibited a similar problem with its orbit, which was solved by a French mathematician named Urban Le Verrier. Le Verrier suggested Uranus's erratic orbit could be explained if the gravity of another, more distant, planet was pulling at it. To make a long story short, Le Verrier calculated where such a mystery planet should be, told some astronomers where to look, and one of them pointed his telescope toward that region of the sky.

They found Neptune.

Decades later, many astronomers expected Mercury's wobbly orbit would be explained the same way: the discovery of a new planet, or maybe host of small planets. In the end, of course, no such explanation could be concluded—because there is no Vulcan. In 1915, Albert Einstein, in his lectures on general relativity, explained away the need for Vulcan. The Sun's massive gravity curved space-time, and that is what caused Mercury's unusual precession. How Einstein came to this solution is a fascinating story in itself, but this is not the place for it. (See the Further Reading section at the end of this book for my recommendations on the Vulcan story.)

And anyway, all that was still to come. Einstein wasn't even born until 1879, so Vulcan was still very much a viable theory in 1878 (although a theory, admitted Rochester astronomer Lewis Swift, "whose very existence astronomers generally were inclined to doubt.") If it *did* exist, however, the eclipse of July 29th was thought to be the perfect opportunity to try to find it. Like big game hunters, a few of the astronomers heading to the Rockies hoped to bag themselves a new planet—"a planet," Swift said, which to that time had been "seen only by the eye of faith." Their hunt for Vulcan, one of the great unsolved problems of astronomy, was described eloquently in the *Washington Post*:

> Advantage will be taken of the Sun's obscuration by the dark moon to examine the heavens in the neighborhood of the Sun for one or more small bodies that shall be added to our solar universe. Besides this careful telescopic scrutiny, large photographs will be taken of the heavens, in order to allow the small planets, if any there be, to print themselves upon the photographic plate, and thus announce to the world the fact of their existence and their influence upon the motions of their neighbor—Mercury. If any of our astronomers should be so fortunate as to discover any such bodies, his discovery will be heralded throughout the scientific world, and he will be ranked with Le Verrier, the discoverer of Neptune, and Asaph Hall, the discoverer of the satellites of Mars.

Two astronomers in Wyoming were the first to spend the few precious minutes of totality looking for Vulcan. They were James C. Watson (director of the Detroit Observatory in Ann Arbor, Michigan) and Simon Newcomb (director of the Nautical Almanac Office). Watson was a large and affable man. Weighing in at some two hundred and forty pounds, his students affectionately called him "Tubby." Newcomb, in contrast, was brilliant but dour, the sort of person who commanded respect.

Astronomers in Rawlins, Wyoming, including the following: Simon Newcomb (far left); James C. Watson (6th from right); Henry and Mary Anna Draper (3rd and 4th from right); Thomas Edison (2nd from right); Norman Lockyer (far right). Courtesy Carbon County Museum.

Newcomb initially planned to observe from Creston, Wyoming. But Creston was frightfully small, and even more frightfully windy. It had only sixteen inhabitants and a cluster of rickety buildings, all of which existed to support the Union Pacific Railroad. So the decision was made to move to nearby Separation, on the Continental Divide. There, a large dune would provide shelter, and soldiers from the US Army erected wooden fencing to give further protection from the wind.

Watson, meanwhile, intended to observe from the much-better appointed Rawlins, a few miles away, with the large group of astronomers including Draper and Lockyer, as well as Edison. But on the morning of the eclipse, he made a short trip with Lockyer to visit Newcomb, and it was decided that, because both Watson and Newcomb planned to hunt for Vulcan that afternoon, they should do so together.

Hunkered behind Newcomb's dune, they endured hours of hot wind that blew sand all over them and their telescopes. Anxious and worried, all they could do was wait ...and hope. Miraculously, just as the eclipse began, the wind died down and the dust cleared from the sky.

At the moment of totality, Watson immediately began scanning for Vulcan. Newcomb, however, was so overcome with the sight of the corona that he wasted a full minute simply gawking at it. The minutes sped by, and Watson thought he'd found two potential candidates for Vulcan. He marked their locations on his chart and rushed over to Newcomb, who was now at his telescope, scanning furiously. But Newcomb was in the middle of his own measurement, and couldn't double check Watson's findings.

And then, all too soon, totality was over.

Over the next ten minutes, as the eclipse shadow raced south through Colorado, other astronomers tried—and failed—to find Vulcan. Edward Holden, after observing from a hotel roof in Central City, reluctantly telegraphed, "I find no Vulcan as large as sixth magnitude." Elias Colbert, in Denver, had the same result. And finally, Asaph Hall, at La Junta, admitted defeat: "I saw nothing but the stars."

Newcomb had planned on one final failsafe to catch the evasive planet. Newcomb's assistant at the Nautical Almanac Office, David Peck Todd, was stationed at Dallas. They had arranged for a Union Pacific telegraph operator, standing by in Wyoming, to telegraph the position of any potential Vulcan sightings to a station in Texas near Todd. At that station, a rider with a horse was ready to receive the information and rush it to Todd, about a quarter mile distant, who could confirm the sighting. By this inventive use of the telegraph, they hoped to outrun the speeding eclipse by even speedier electric current.

Alas, Newcomb and Watson appeared to have forgotten the plan, and the telegram was never sent. Todd did what he could on his own, but the skies above Dallas were hazy, and he couldn't see much, anyway.

Although no other Vulcan sighting had been confirmed, one other astronomer corroborated Watson's observation. That was Lewis Swift, one of the few faithful believers in Vulcan, who had observed the eclipse from Denver. Swift claimed to have observed the same thing Watson did, the result of a lucky adjustment Swift had made to his telescope at just the right time during totality.

Watson didn't need further confirmation. He was convinced he'd found Vulcan, and that Swift had seen it, too. So he proceeded as if the discovery was certain. The press, always eager for a good story, was happy to oblige him. From Laramie to London, the announcement was made. "Professor Watson ... had taken the job of FINDING VULCAN. And he found it." Astronomer Charles Young, of Princeton, tentatively supported Watson, writing in the *New York Times* that Vulcan had finally been "run down and captured."

However, George Biddell Airy, the Astronomer Royal of England, was quick to question Watson's discovery. Writing in *Nature* on August 8th, he noted that the location of Vulcan as described by Watson was nearly the same as the known position of the distant star θ Cancri, "and that there may be a possibility that [Watson's Vulcan] is in reality this star."

Airy conceded that the magnitude of θ Cancri was different from the magnitude Watson gave for Vulcan, and therefore the debate could still be considered open—although barely. In the scant minutes Watson had to observe his potential new planet, he likely misjudged its magnitude. "The discrepancy," Airy cautioned "may very easily occur in the hurry of such a sensational observation, as on these occasions, the time at the disposal of the observer is so limited."

Of course, we know that Watson did not find Vulcan, because it isn't there. Most of his contemporaries soon doubted that he had, too. Sadly, Watson died of pneumonia a few years after the 1878 eclipse. But the idea of Vulcan lived on, because *something* had to account for Mercury's odd orbit. Occasional Vulcan sightings persisted for almost another four decades, until Einstein finally gave astronomers reason to stop looking in 1915.

In the end, finding Vulcan had been but one of many objectives of the observers during the 1878 eclipse. The day after the eclipse, scientific and popular reporting of the results began in earnest, and continued for months, even years.

PART III

RESULTS

17. Crossing the Charmed Circle

Depending on who you were in 1878, the eclipse was either a very communal or a very exclusive event. For the general public, the eclipse was a natural spectacle open to all. But the astronomers had to maintain a certain distance from the crowds in order to make their observations unmolested, especially during the few crucial minutes of totality. They needed a defined space in which to do their work, and perfect silence to hear the methodic count of the timekeepers calling out the passing seconds. To enforce the separation between the professionals and the public, military and police guards were used. This layer of official protection added to the aura of inaccessibility that surrounded the astronomers. Then, as now, scientists at work were seen as somehow special, different, unique.

The mayor of Denver provided Chicago astronomy professor Elias Colbert with "a police officer to keep order on the ground during the eclipse." And the summit of Pikes Peak, which at the time was a military reservation, was made off-limits to anyone not already approved to be there by the US Signal Service. An announcement in a local newspaper warned off interlopers by reporting: "The government has taken possession of Pike's Peak, around the base of which a guard will be placed with orders to permit none to pass but those having an order from the secretary of war." Of course, ringing the base of Pikes Peak with guards was a practical impossibility. What in fact occurred was that a few soldiers were placed near the summit to give the astronomers up there some freedom. (The tourists still came, by the way, but they kept their distance.) In both of these cases, the message to the public was clear: *keep out*.

However, it is from Maria Mitchell that we get the most eloquent description of the boundary between professional astronomers and the public. "It is always difficult to teach the man of the people that natural phenomena belong as much to him as to scientific people. Camping parties who put up telescopes are always supposed to be corporations with particular privileges, and curious lookers-on gather around and try to enter what they consider a charmed circle."

In the days leading up to the eclipse, the astronomers were more than happy to show off their instruments to locals and tourists. But on the day of the eclipse itself, especially during the few minutes of totality, even a momentary interruption could ruin an observer's concentration, and thereby months of preparation. The seriousness of the situation was made clear in the *Laramie Daily Sentinel*: "An order was issued by the chiefs that if, during the totality, anybody came near or spoke to [the astronomers], 'shoot him on the spot.'" I have to assume (or at least hope) that this order was made in jest, as it seems unlikely—even in the Wild West—that disrupting science would be a capital offense.

After all, except for those crucial minutes of totality, the boundary between astronomers and the public was probably far more fluid than it would have been had the eclipse occurred in the eastern United States, where social hierarchy was far more entrenched. Mitchell had the remarkable experience of setting up her Denver eclipse camp on private land, apparently without first asking permission. Then the owners came along and offered to set up a fence around her camp, to keep out trespassers!

As she put it, "In New York or Boston, if I were to camp on private grounds, I should certainly ask permission. In the far West, you choose your spot of ground, you dig post-holes and you pitch tents, and you set up telescopes and inhabit the land; and then the owner of the land comes to you, and asks if he may not put up a fence for you, to keep off intruders."

No fence was erected, but Mitchell did hire someone to "send away the few strollers who approached" during the eclipse. However, the existence of the "charmed circle" does not mean that the public had no role in making useful eclipse observations. In fact, many members of the public were invited by the professionals to participate. Crossing the charmed circle was possible, but it required some training first.

In Denver, Professor Colbert gave instruction on sketching the corona to a few interested members of the public. These trainees, reported the *Denver Daily Tribune*, "will make drawings of the Sun at the time of the eclipse." Colbert trained sixteen assistants, each of who was assigned to draw one part of the corona. Why did he do this? Because mounting an expedition to the Rockies was costly, and not all astronomers could bring along a team of assistants. So, in some cases, members of the public were deputized and given a crash course, to make up for the lack of an astronomer's usual helpers.

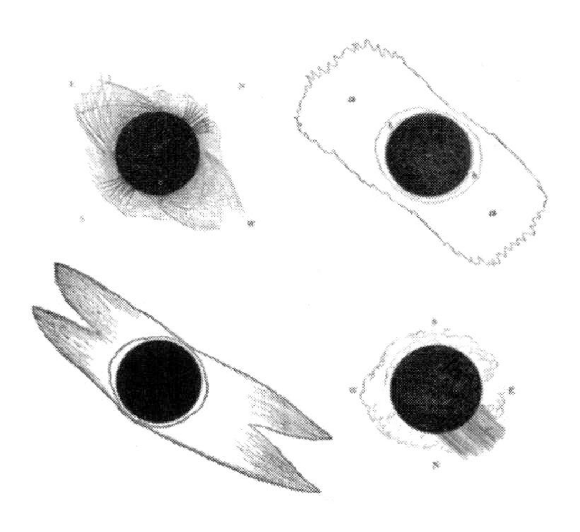

Corona sketches, made by both professionals and amateurs during totality, using the template provided by the US Naval Observatory.

After the eclipse, public reports and drawings flooded Harvard College Observatory and Washington DC. In a few cases, the public's observations were better than the professionals', such as the case of a studio photographer from Central City, Colorado, who took a photo of the corona with the same camera he used to make portraits. In the end, his photo of the corona was deemed more successful than many of the eclipse photographs taken by the professional astronomers, who had been equipped with specially outfitted scientific cameras from the Naval Observatory.

Public contributions, it turned out, were not only desirable, but actually useful, leading one sarcastic editorial to quip that it was about time the professionals shared the limelight with the public. "For hundreds of years, these astronomers have been loafing around in large numbers at every eclipse, always taking the front seats and sticking their huge telescopes up so as to hide the performance from everybody else."

* * *

The eclipse of 1878 was probably the first time that eclipse tourism became a mass phenomenon in America, with thousands of people traveling thousands of miles to reach the path of totality—a trend that continues today. (In fact, it is estimated that tens of millions of Americans will travel hundreds—even thousands—of miles to see the eclipse of August 21, 2017.) It may also have been the first time the public was asked, on such a grand scale, to participate in the scientific observations of a total solar eclipse. The US

Naval Observatory circular provided everyone with basic instructions, and some of the visiting astronomers trained a volunteer corps of deputy observers. It's no wonder frontier citizens felt like they were "among the favored mortals of Earth." And like eclipse tourism, public participation in astronomical investigations also continues—from amateur astronomers making important discoveries, like David Levy, co-discoverer of Comet Shoemaker-Levy 9, to regular people donating time on their home computers to organizations like SETI—the Search for Extra-Terrestrial Intelligence. The opportunities for the American public's participation in astronomical research are endless, and this phenomenon began in earnest in 1878. But there was one group whose aid was not utilized during that eclipse. In fact, they were even ridiculed, because of an unfortunate and extreme prejudice against them. These were the Native Americans, and the bias against them is perhaps the low point to this story. But it is important to tell it.

18. The Savage and the Scholar

Unfortunately, the response of Native Americans to the 1878 eclipse is only known secondhand. A firsthand account of the eclipse from the Indian perspective, whether from an interview or written record, might have helped mitigate the inevitable prejudices of the whites.

Historian Alex Pang, who has written widely on nineteenth-century solar eclipse expeditions, notes: "There are no sources that describe the experience of eclipse-watching from [the indigenous peoples'] point of view. This is especially unfortunate, for a comparison of the experiences of watching an eclipse—and of watching members of another culture watch an eclipse—would have been interesting. As it is, we can look only one way, from European [and American] to native."

The few available reports of the Native American response to the 1878 eclipse were written by white Americans. They reveal the common, late-nineteenth-century view that the Native Americans were "savages." I suppose the ongoing tension on the frontier in the late 1870s made such a bias inevitable. These reports imply that Native Americans were incapable of understanding the eclipse as anything other than a supernatural phenomenon.

Consider the story told by Hugh Lenox Scott. At the time, Scott was a frontier cavalryman, but would later be superintendent of West Point, and eventually Chief of Staff of the US Army during the first part of World War I. Scott was tough as nails and highly intelligent, and learned much about the various plains tribes' ways of life, even developing remarkable proficiency in their languages. During the eclipse of 1878, Scott was camped near the Cheyenne Indians in the Dakota Territory at Bear Butte (now a South Dakota state park, near the town of Sturgis). The Cheyenne were being held as prisoners of war, and Scott spent his time among them learning as much as he could. Although Bear Butte was outside the path of totality, they would have seen a near-total solar eclipse. Scott recalled the reaction of the Cheyenne as follows:

"Our camp was immediately under and north of Bear Butte, a single peak some miles east of the Black Hills range, which was regarded by the Cheyenne as their principal medicine place. Many of them climbed to the summit, where they left presents to the medicine that inhabited the mountain. While the Cheyenne were still in that camp, an eclipse of the Sun took place that was announced by the public press. I told the Cheyenne to expect it several days beforehand, but they did not believe me. They became very much excited when the eclipse began, shooting off guns and making every sort of noise they could to frighten away the evil medicine which they thought was destroying the Sun. Their treatment was highly successful—the Sun recovered."

Someone who saw the eclipse from Taos, New Mexico (also outside the path of totality), reported that the Pueblo Indians had a similar response to the eclipse. "As the light began to grow dimmer, they became confused and frightened, and began to run hither and thither, and to gather in groups. The darker it grew, the more excited they became." The report goes on to describe how their chief made his people do penance by running races, and stoke their cooking fires until they were blazing. In this manner, they were able to appease their god "to desist in his punishment," and the Sun eventually reappeared.

Yet another report, from an observer at Fort Sill, Oklahoma, wrote that the eclipse "frightened the Indians badly. Some of them threw themselves upon their knees and invoked the Diving blessing; others flung themselves flat on the ground, face downward; others cried and yelled in frantic excitement and terror." The Indians were only mollified when an elder came out of his lodge with a pistol and fired it at the

Sun. The shot coincided with third contact, leading the Indians to assume the bullet had driven away the moon's shadow.

Somewhat more charitably, though still condescendingly, Maria Mitchell referred to the Native Americans as the "untutored Indian." Her phrasing at least acknowledges that the Native Americans *could* have understood the scientific explanation of an eclipse, crediting them with intellectual potential that others seemed to deny them. Perhaps Mitchell was so used to defending a woman's ability to learn as well as a man could, that it was natural for her to assume the Indians were equally capable, given the right opportunity. Nevertheless, she too supposed that the Native Americans thought of the moon's shadow as the displeasure of their god. "*We* saw the giant shadow as it *left* us and passed away over the lands of the untutored Indian; the savage, to whom it is the frowning of the Great Spirit, is awe-struck and alarmed; the scholar, to whom it is a token of the inviolability of law, is serious and reverent."

White Americans' anxiety about Indian hostility is certainly one explanation for dismissing them as "savages." Christian Peters, astronomer at Hamilton College, New York, was invited to observe the eclipse from Montana, but declined. He wrote to US Naval Observatory astronomer Edward Holden, "It is a great temptation, but I ought not to go. So, you go to Montana. Take care of not being scalped by the Indians!"

Peters's quip masked a legitimate concern. Just two years earlier, George Armstrong Custer and his regiment had been crushed at Little Big Horn. Tensions were high across the Great Plains. In the summer of 1878, unrest among the Utes in southern Wyoming and northern Colorado made otherwise advantageous areas in the eclipse path unsafe for observations. This anxiety even resulted in the astronomers' penchant for describing the success of their observations in terms of the whites' biggest frontier fears: being scalped. As Norman Lockyer joked about James Watson's reported discovery of Vulcan, "Prof. Watson's belt, as the papers here put it, is graced with the scalps of I know not how many minor planets." When words like that start to color the language of scientists, even in jest, we can be certain that a certain degree of angst has crept in. After all, as Pang notes, white accounts of native responses to the eclipse can often "reveal much more about European [and American] anxieties than actual native behavior."

After all, ancient North Americans may have been keen eclipse observers. As astronomer Tyler Nordgren notes, in his recent book *Sun Moon Earth*, the ancient Anasazi peoples drew what looks like a very accurate pictograph of the solar eclipse of July 11, 1097 in New Mexico's Chaco Canyon. It may have been a simple observation set down on stone, or a more ominous cultural warning. Whatever the significance of that eclipse to the Anasazi nine hundred years ago, we know that eclipses have been seen as omens of doom for many cultures, including European and American Christian denominations. Even in 1878.

And so there is a certain irony in the dismissal of the Native American's response to the eclipse as

"The enterprise of one of our businessmen, knowing that the Millerites would expect the world to come to an end ..." Colorado Springs Weekly Gazette, August 3, 1878.

superstitious. After all, members of the religious sect of Millerites in Colorado Springs believed that the 1878 eclipse signaled the end of the world. According to the local paper, one enterprising citizen walked among them just before the start of the eclipse, helping them make their final preparations. Tongue-in-cheek, the paper noted how the man carried around a "sample coffin under his arm, and took a large number of orders at a twenty-five percent discount."

It would've been nice if the 1878 eclipse had been an omen of peace and cultural understanding between whites and Native Americans. It was not. That would only come with time, and, unfortunately, more bloodshed. At least, however, the eclipse was an event that brought astronomers, tourists, and frontier residents together for the advancement of science, and in a spirit of mutual good will.

19. An Exceedingly Pleasant Memory

In the days following the eclipse, all of the anticipation—to say nothing of the anxiety—that had been building for months was converted into a sort of victorious energy. The event had been a huge success, largely due to the good weather. Self-congratulation was the order of the day, as frontier newspapers praised themselves, their readers, and pretty much everyone and everything else in the Rocky Mountains.

The word "great" was added in front of "eclipse" nearly everywhere, from major newspapers to the article that appeared in *Harper's Weekly*, with the stunning front cover showing tourists viewing the eclipse from high in the mountains. Although the exact phrasing varied, everyone agreed.

They had accomplished "The Great Solar Eclipse."

The rest of America (for whom the electric telegraph was the closest thing the late nineteenth century had to the Internet) wanted news of the eclipse, and they wanted it *now*. Frontier papers, especially those in Colorado, began publishing "eclipse editions."

The *Denver Daily Tribune* sold six thousand copies of its eclipse edition on July 30th, noting "the demand was almost if not quite unprecedented in the history of Colorado journalism." On July 31st, demand continued. "Another large 'eclipse edition' was printed in response to telegraphic orders from different parts of the country, swelling the total number of papers distributed to nearly *ten thousand copies*. ...Almost every citizen preserved one for future reference."

The eclipse had put local pride on the line. Frontier towns and their newspapers immediately started bickering over who got the best view of the eclipse, who had the best turnout, whose scientists got the best results, and who generally did the eclipse *better* than the others. And no claim to victory went unchallenged. When the *Denver News* ran the seemingly innocuous headline "Denver Eclipse," Georgetown's *Colorado Miner* nearly flipped:

"Rather cheeky use of the word Denver as an adjective in this connection! Had it read 'Denver Eclipsed,' it would have come nearer the truth, for the view from Argentine and Grays Peak was far superior to that of our friends in the capital."

THE ECLIPSE!

A GLORIOUS DAY FOR COLORADO!

Our State Doesn't "Go Back" on Herself.

THE GREAT PHENOMENON FROM VARIOUS POINTS.

IN GEORGETOWN.

Last Monday dawned clear and glorious. Not a cloud obscured or flecked the azure of the concave dome that spanned the heavens from mountain top to mountain top. The day wore on and the shining orb wheeled along on its path of splendor. The hour for "first contact"

From the Georgetown Colorado Miner, August 3, 1878.

Other papers boasted about the results obtained by the astronomers in their area, including the possible discovery of Vulcan or the measurement of the giant streamers extending from the Sun's corona. In that competition, the *Colorado Springs Gazette* may have won top prize for scientific reporting, as it was able to procure an exclusive from Cleveland Abbe. Abbe, having observed from the lower elevation of 10,000 feet after being rushed off the Peak due to his altitude sickness, had recovered enough to observe and sketch the streamers.

Many of the astronomers believed the streamers were extensions of the solar atmosphere. But not Abbe. He proposed that the giant rays "had nothing to do with the observer, or [Earth's] atmosphere, or moon, no, not even with the Sun," but were in fact "grand streams of meteors rushing along in parallel orbits about the Sun." It was a novel theory, and the *Gazette* gleefully wrote that Abbe's "observations will probably result in the overthrow of all previously entertained theories respecting the character and cause of these streams of light." *Yes*, the *Gazette* all but printed: *You read it here first, folks!*

In Wyoming, the *Laramie Weekly Sentinel* had a report titled "The Great Eclipse," which eagerly discussed Watson's discovery of Vulcan, the streamers, and Edison's tasimeter, marveling: "All this work, and all those discoveries, had to be done in less than three minutes. What momentous interest hung upon this brief space of time!" Meanwhile, Texas praised the discipline of their astronomers. The *Galveston Weekly News* of July 30th wrote that the observers behaved like old veterans in "a decisive moment of battle, when all are trained to duty at their respective posts."

The astronomers were described in military terms elsewhere, too. The *Boulder County News* likened them to a battalion of sages following a guiding star. "Everybody was glad on account of the wise men from the East who had come to plant their telescopic artillery against the Colorado sky." And the *Colorado Transcript* called them soldiers of progress, who "came to Colorado ...bringing with them the pacific weapons of civilization and science."

Some of those astronomical "weapons" were left behind, helping Colorado become a new center for astronomy. Colorado College, founded in 1874 in Colorado Springs, obtained a four-foot telescope from New York. That telescope had been brought to

Colorado with Darwin Eaton, from Brooklyn's Packer Institute, who observed the eclipse from the mountain town of Idaho Springs. Colorado College was able to purchase Eaton's telescope after local donors raised enough money to buy it before Eaton returned home. It may have been Colorado College's very first scientific instrument, and it began the astronomical career of Frank Herbert Loud, the school's professor of mathematics-turned-astronomy. Loud soon helped Harvard College Observatory's director, Edward Pickering, explore Colorado for potential high-altitude observing sites. Loud also taught a young Colorado student named Donald H. Menzel, who would eventually hold Pickering's job as director of the Harvard College Observatory. In 1940, Menzel established the High Altitude Observatory near Climax, Colorado. It took a few decades, but Colorado went from providing world-class observing sites, to eventually producing world-class astronomers.

But all that was in the future. Immediately after the 1878 eclipse, Colorado was ready to turn its sunny skies and clear atmosphere into a new business. "The world should know that fair weather, good health, eclipses, transits of Venus, and such, are kept for sale in Colorado, forming the third great industry in the State," deadpanned one paper. Another was simply proud: "None of us were ashamed of the behavior of the heavens on that important occasion. Until mid-day, not a speck of a cloud appeared. It was an honor to our climate."

Despite the celebrations, it was time for the astronomers to return home. As they boarded their trains, the residents of the frontier cheered them like departing heroes.

"Such an array of distinguished scholars," waxed the *Laramie Weekly Sentinel,* "furnished a rare opportunity to us frontier residents to enjoy some of the wonders of science; and while they were nightly engaged in scanning the heavens with their powerful glasses, the citizens—old and young; men and women—flocked about their tents with eager curiosity, and they never tired of showing and explaining to the citizens the use of the instruments, and showing them the wonders of the heavens through their glasses."

The report concluded, "The people of Rawlins will never tire of talking of their distinguished visitors ... and the admiration seems to have been mutual, too, for we heard all the different members of the party speak in high terms of the people and of the pleasantness of their stay with them."

The feeling was indeed mutual. Henry Draper, leader of the Rawlins expedition, gratefully thanked the town. "Of the citizens of Rawlins, it is only necessary to say that we never even put the lock on the door of the observatory, and not a thing was disturbed or misplaced during our ten days of residence, though we had many visitors."

To this, Edison's friend George Barker added, "The agreeable party, the pleasant surroundings, the charming weather, the kindness of friends, and above all, the capital success of the observations, made the Draper Eclipse Expedition an exceedingly pleasant memory to us all."

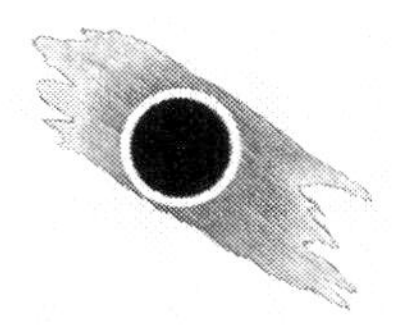

20. The Generation Now on Stage

he popular press published their eclipse reports almost immediately. But the astronomers' reports took weeks to appear, and some were not published for months, or even years.

The English astronomers announced their results in British journals like the *Astronomical Register*, *The Observatory*, and the *Monthly Notices of the Royal Astronomical Society*. In 1878, however, America did not have its own dedicated astronomy journals. That changed in the 1880s and 1890s, with the founding of *Popular Astronomy*, *Astronomy and Astrophysics*, *The Starry Messenger*, and the reappearance of the *Astronomical Journal*.

Until then, the American astronomers' eclipse reports found homes in more general academic publications. Charles Young, returning to New Jersey from Denver, published an extensive account of the eclipse in his university's journal, *The Princeton Review*.

Most notably, he estimated that there had been hundreds of astronomers at the eclipse, including many amateurs. "My list," he wrote, "does not include a host of independent observers, who, singly or by twos and threes, were scattered all along the line, many of them competent, skillful, and well equipped for their special work. It would probably be quite within the mark to set down the total number of scientific observers as at least two hundred."

But Young published hastily and believed that Vulcan (or perhaps multiple Vulcans) had, indeed, been found, lauding "the brilliant discovery which will always give the eclipse of 1878 a place in the annals of astronomy—the discovery of one, and perhaps two, intra-Mercurial planets by Professor Watson and Mr. Swift."

Young knew his report was preliminary, relying, as he did, "upon mere newspaper reports, which are always more or less inaccurate." Nevertheless, he also knew that something important had happened in those three minutes of totality above the Rocky Mountains. "It is evident," he wrote, "that the recent eclipse has materially advanced our knowledge of the Sun and his surroundings." This new knowledge included data linking the variability of gasses in the corona to sunspot cycles, which were known to have a period of maximum and minimum density that varied over a period of years. Photographs taken of the corona represented the application of a new technology to an ancient science.

The German-born British physicist Arthur Schuster was stationed at Las Animas, in Colorado. He was only twenty-six in 1878, but went on to have an extremely distinguished scientific career. He, too, recognized that the 1878 eclipse was the beginning of something big, not just for astronomy, but for *American* astronomy: "We have arrived at a stage of eclipse observation where regular work may be said to begin. The most important results, however, will only come out by a study of minute changes in different parts of the corona during different eclipses. Some of the American observers during the present eclipse (Prof. [John] Eastman in his spectroscopic observations, and Prof. Arthur Wright in his polariscopic observations) have been alive to this fact, and may be said to have inaugurated a new era."

"Inaugurated" is the appropriate term, in so far as the 1878 eclipse was an early moment in history of the new astronomy. In fact, in the final analysis, the overall scientific results were less than spectacular. Vulcan had not been found. Edison and his tasimeter did detect coronal heat, but so had an Italian astronomer almost forty years before, and with far less sophisticated apparatus. Photographs of the corona were decent, especially those made by Schuster, but most failed to capture the massive streamers—the most visually arresting part of the eclipse. (Schuster made a major breakthrough during the 1882 eclipse, in Egypt, when he captured the first-ever photograph of the corona's spectrum.) And the spectroscopic data—for those who were able to collect any—was often inconclusive, leading more than one of the astronomers to doubt the corona had "any considerable mass of incandescent gas or vapor" at all, and to speculate that the streamers were actually *external* to the Sun, not part of it. The one important exception to the new spectroscopic data was the surprising fact that the corona contained quite a bit of the element calcium, but even then, astronomers concluded that the 1878 eclipse "added very little to our knowledge of the coronal spectrum." But at least they had begun employing those new techniques, and they would improve them with each new eclipse observed.

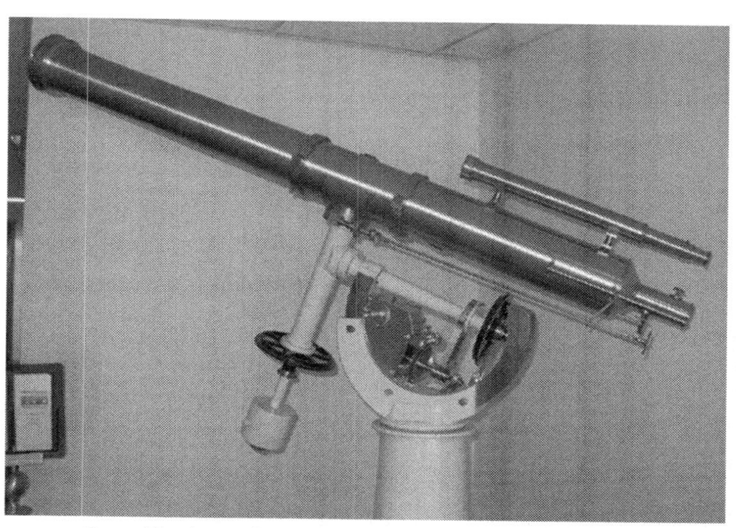

One of the heavy brass Alvan Clark telescopes the US Naval Observatory loaned to the 1878 eclipse expeditions. Courtesy US Naval Observatory Library.

"With this eclipse," wrote Mabel Todd, who had observed with her astronomer-husband David Todd from Texas, "the scientific world was fully launched upon the most approved modern methods of attacking the great problems of the Sun, many of which must wait years for these fleeting minutes to afford the necessary opportunity for their complete solution."

And so, the astronomers would have to keep investigating. That would mean waiting for the next total solar eclipse ...and the next. They wouldn't have to wait too long; eclipses occur roughly every two years, although not always in places as accessible as the American frontier. Some would only be visible in the middle of the ocean, or in even more inhospitable locations like the Arctic.

Ten years later, Samuel Langley agreed that the 1878 eclipse and the ones that followed had been a series of baby steps. "There have been other eclipses since; but in spite of all, our knowledge of the corona remains very incomplete, and if the most learned in such matters were asked what the corona was, he could probably answer truthfully, 'I don't know.'"

After all, with totality lasting maybe three minutes, it would take a dedicated astronomer forty years of chasing eclipses all over the world—never missing one, never having cloudy weather, never having their equipment fail—to get even an hour's worth of coronal observations. Clearly, advancements in solar physics were going to take some time.

* * *

The definitive collection of the Americans' eclipse results did not appear until 1880. It was given the rather unexciting title, *Reports on the Total Solar Eclipses of July 29, 1878, and January 11, 1880, Issued by the United States Naval Observatory*. Although it tacked on some reports from an eclipse that occurred in 1880, those reports came to less than twenty pages. The bulk of the book, nearly four hundred pages, was devoted to the 1878 eclipse.

It is a large book, dull and featureless. You'd never pull it down off of a shelf based on how it looks. I know, because I own a copy—it is sitting on my desk as I write these words, and it is one of the least exciting-looking books in my library. Frayed and faded, its cover may once have been dark blue, but is now a rather uninspiring shade of green-gray. The binding is coming off, and the pages shed little flakes when I turn them.

But inside, it is utterly fascinating.

Within the dense text and equations, all printed on thick paper that has been browned with age and smells of dust, are the official 1878 eclipse reports, from Creston to Dallas. They include observational reports; spectroscopic and polariscopic data; records of time, temperature, longitude, and latitude; discussions about what investigations astronomers should make during future eclipses; and, at the end, dozens of sketches of the corona during totality.

The results were not revolutionary. But taken together, all that information tells the story of a science on the cusp of a new era, and to me, that's just as exciting. Most of the scientific revolutions we hear about, like the Copernican revolution or the Darwinian revolution, were merely preliminary. After any revolution comes a period of rebuilding, or—in this case of the new astronomy—of building up. And that's when some of the most interesting scientific work gets done. The American astronomers who witnessed the eclipse of 1878 were the first generation of the "new" physics-minded astronomers: the "generation now on the stage," Samuel Langley called them.

Although 1878 was not the very beginning of the new astronomy, the eclipse *was* the start of something equally significant: the beginning of high-altitude astronomy. In the next and final chapter, I'll explain why.

21. The Beginnings of High-Altitude Astronomy

In the summer of 2009, I found myself in a stiflingly hot seminar room in Budapest, Hungary. I was giving a talk at an international history of science conference. Europe was in the middle of a heat wave, and the old building I was in did not have modern air conditioning—or even windows that opened more than a crack. As beautiful as Budapest is, I was already looking forward to getting back to the cooler air of Colorado, and its mountains.

Despite the sweltering heat, I finished my talk. In that presentation, I had proposed that the 1878 eclipse was the beginnings of high-altitude astronomy. I discussed the arguments the American astronomers made for, and against, putting telescopes at higher elevations.

The feedback from my audience was encouraging. Since then, I've scoured the reports that came out of the eclipse, and I remain convinced that 1878 was, indeed, the year when a critical mass of working astronomers realized that placing telescopes on mountains was a rather good idea, and should be pursued.

But not all astronomers felt this way, and in fact, quite a few were actively against it.

It may seem strange that some astronomers in 1878—many of them leaders in their field—would have doubted the advantages of putting telescopes on mountaintops. After all, it's a tenet of modern astronomy that telescopes placed in the thinner atmosphere of higher elevations are, all other conditions being equal, better able to see into space. The billions of research dollars that have been invested in high-altitude observatories over the past century attest to this.

Yet it was not always self-evident. It would take a few hundred astronomers, gathered on the high western plains and Rocky Mountains, to begin to make the case.

The meteorologist Cleveland Abbe, who had observed from Pikes Peak, noted that one of the goals of astronomers in 1878 was to determine "the advantage of elevation above sea-level as increasing the visibility of stars." William Harkness, who had observed in Creston, later recalled that there was still no consensus. "Of late, there has been much discussion as to the relative advantages of sites at sea-level, and upon high plateaus, for observatories."

Not every astronomer who visited the Rockies that summer had these concerns in mind. Many were just there to see the eclipse, write their reports, and go home. Others were happy to return to observing at lower elevations after spending weeks suffering through chilly nights in the thinner air, and days spent fighting the wind and weather.

Harkness found that elevation alone did not necessarily improve clarity. In Wyoming, the thinner atmosphere was offset by high winds and extreme temperatures. "It is impossible to avoid the conclusion," he wrote, "that in an apparently clear atmosphere, superior altitude does not necessarily ensure superior transparency."

Another observer in Creston, the Frenchman Etienne Trouvelot, agreed wholeheartedly with Harkness. The hot, dusty plains of Wyoming were, for Trouvelot, simply awful. He much preferred observing on the east coast of the United States. "I had anticipated an unusually fine opportunity for astronomical observations at Creston, at an altitude of above 7,000 feet, and under the pure sky of a mountainous region. But in this, I was greatly disappointed, as during the seven days that I observed the sky with the telescope, I did not have a single observation which was satisfactory."

And yet, Harkness and Trouvelot were in the minority. Most of the astronomers who commented on the altitude question believed that observing from higher elevations was beneficial. Perhaps the most significant among these was the telescope maker Alvan G. Clark. His company made some of the best, and largest, telescopes in the world, and their equatorial telescopes were the same ones loaned to most of the 1878 eclipse expeditions by the US Naval Observatory, including the one Langley dragged to the top of Pikes Peak.

Clark was not just a great telescope maker; he was also an accomplished astronomer, having won the Lelande Medal from the French Academy of Sciences in 1862 for his discovery of the companion star to Sirius. So it is highly significant that, after observing the 1878 eclipse in Creston—and from the very same location as Harkness and Trouvelot—Clark made the rather astonishing claim that, in his professional opinion, a telescope with a small, three-inch objective lens in Wyoming was equal to a much larger telescope with a twelve-inch objective lens at sea level. In other words, higher elevations made telescopes much more powerful.

Other observers of the 1878 eclipse made similar claims. Norman Lockyer stated it was far easier to study the solar corona from a higher altitude, like Pikes Peak, where the corona remained visible a full five minutes after totality ended. John Eastman, observing from Las Animas, not only saw the moons of Jupiter with his naked eye, but he also claimed that "there seemed to be in Colorado an increase of about one unit in the magnitude of stars."

In La Junta, Asaph Hall claimed that Colorado enabled naked-eye observations that were simply impossible in Washington, DC. The "elevated plains offer advantages for astronomical observations that have not hitherto been made use of. ...They are now easy to access, and an experienced astronomer with a good telescope should be placed in an elevated position for several years."

And Cleveland Abbe reported that the higher elevations of Pikes Peak provided an "excellent naked-eye view ...the dark spaces of the Milky Way seemed to entirely disappear, being replaced by a multitude of fine stars and hazy light ...the whole *via lactea* assumed a continuity and a body that I never saw at lower altitudes."

* * *

Why does all this matter? So what if some astronomers in 1878 thought that higher altitudes provided better observing conditions, and others didn't? Surely, the "altitude question," as I call it, would have been resolved at some point, Rocky Mountain eclipse or not.

Perhaps. But in 1878, American astronomy was at a turning point. Earlier that year, in Washington DC, a politically charged debate had been raging over the relocation of the US Naval Observatory. One of the most contentious elements of the debate was whether or not the Naval Observatory should remain under the control of the Navy, or be turned into a national astronomical observatory with a civilian director. It is within these debates we find the first public references to the "altitude question."

The *New York Times*, which had been reporting on the debates over the relocation of the Naval

Observatory at the end of 1877, and into the first part of 1878 (well before the July 29 eclipse), mentioned the "proposition proceeding from various scientific men to move the observatory from the capital. No particular place has been advocated, but one idea is to go up to the top of the Allegheny Mountains to avoid the impurities of the lower stratum of atmosphere, which some think affects the accuracy of the observations."

Then, according to historian Steve Dick, as astronomers and politicians continued to fight over the costs and merits of relocating the Naval Observatory, there was a rival resolution introduced in the Senate for a commission of scientists to undertake a study of the "establishment and location of a national observatory at high altitude on the Western plains." (This could explain why Harkness, an employee of the Naval Observatory, seemed unenthusiastic about high-altitude observing.)

And then there was the matter of funding. At the time, funding was still going toward the "old" astronomy. As Samuel Langley lamented: "It is not generally understood that not only the support of the Government, but every new private benefaction, is devoted to 'the Old' Astronomy, which is munificently endowed already; while 'the New,' so fruitful in results of interest and importance, struggles almost unaided."

Therefore, not only was the 1878 eclipse important for the *new* astronomy, it was also a battleground over the changing direction in *American* astronomy. Many of those astronomers were pushing for government support for astronomy beyond its traditional military and commercial applications. In short, they were fighting to modernize their science.

In the end, a government-funded national observatory was not established in the American West, nor did the Navy relinquish control of the new Naval Observatory (which was completed in 1893 in its current location, called "Observatory Circle," which is well within the boundaries of the District of Columbia) to civilian astronomers. Some concession was made to the astronomers, however, as the 1879 *Report of the Commission on Site for Naval Observatory* reveals: "In constructing these buildings, a National Observatory should be provided, which, while satisfying the practical astronomical exigencies of the military and commercial marine of the United States, shall also meet the higher and more universal demands of science, by equality in all its material means with other great national observatories."

That was all well and good, but it still did not put a new observatory any higher above sea level. So astronomers took matters into their own hands. In the 1870s and 1880s, their science was taking a radical turn, emphasizing pure physical research on our Sun and the distant stars. That meant the new astronomy needed observatories in places other than at sea level where the Earth's atmosphere was an impediment. America's High Plains and even higher Rocky Mountains, for the first time accessible by train, were where American astronomy's future would be found.

Edward C. Pickering, director of the Harvard College Observatory, threw down the gauntlet in his influential article "Mountain Observatories," in which he argued that regardless of the outcome, the altitude question needed to be answered. "Much attention has recently been directed to the question whether the conditions are more favorable to astronomical observations on the summit of a lofty mountain than at the level of the sea. The question should, however, be decided . . . valuable results would be obtained, even if the observations in the observatories on the mountain summit were not much better than those at a less altitude."

It was the 1878 eclipse that allowed those astronomers to take their first tentative steps toward those mountain observatories. As far as the astronomers were concerned, it was going to happen whether the government got involved or not. Which is why, within a few decades of the 1878 eclipse, a number of prominent high-altitude observatories were built in the Rockies and Sierra Nevadas, often with private funds. Ten years after the 1878 eclipse, Colorado businessman Humphrey Chamberlin pledged $50,000—a massive sum at the time—to build an observatory for the University of Denver, with a nineteen-foot telescope that was then one of the largest in the world.

Other observatories appeared, perhaps most notably the Lick Observatory on Mount Hamilton in California, which was the world's first permanent mountaintop observatory. Percival Lowell, a wealthy Bostonian who was convinced that there were canals on Mars, established the Lowell Observatory in Flagstaff, Arizona, in 1894. These are but a few of the prominent, early western observatories.

The twentieth century saw the construction of high-altitude observatories all over the country (and the world, for that matter). From the Mauna Kea Observatories in Hawaii, to Kitt Peak National Observatory in Arizona, to the High Altitude Observatory in Colorado (mentioned earlier), the list is impressively long. The University of Wyoming Infrared Observatory (WIRO) is one of the world's most important infrared observatories. It sits on a mountain peak

Samuel Pierpont Langley.

nearly ten thousand feet above sea level, just a hundred miles from Rawlins, Creston, and the other 1878

Wyoming expedition camps. WIRO became operational in 1977—ninety-nine years after the 1878 eclipse. And even today, an effort exists to establish a permanent, remotely operated telescope on the summit of Pikes Peak, where Langley and his team suffered so stoically.

Each new high-altitude observatory that appeared underlined the claims of the astronomers of the 1878 eclipse: in the new astronomy, topography mattered. As the decades went by, and as techniques and technologies improved, telescopes and other instruments at higher elevations provided ever-better astronomical results than those at sea level.

But success did not come overnight. High-altitude observing was a process of trial and error. Langley never forgot how hard it was on the top of Pikes Peak. He advised future observers to make their camps a little lower, where the oxygen is more plentiful and trees offer shelter. "It is desirable to place one of the great telescopes now made in a station far above our lower atmosphere, and the conditions of vision on mountain summits are so increasingly important. We have found at a little over 10,000 feet great purity of atmosphere, but at 14,100 feet, the physiological conditions were at all times such as to be unfavorable to good observation."

Even though the altitude and weather on Pikes Peak nearly did him in, Langley continued to promote astronomical observations from higher elevations. Within a few years, he led an expedition to the even higher summit of Mt. Whitney in California, to study the atmospheric absorption of solar heat. He had become a believer in mountain observatories. Like other veterans of the 1878 eclipse, Langley took what he had learned and applied it to future high-altitude observations.

Therefore, if I were to pick a date for the beginning of high-altitude astronomy, it would be July 29, 1878. Everything before then was speculation on the advantages of higher altitudes, and everything afterward saw astronomers experimenting as they determined the best elevations and locations for the observatories, telescopes, and other instruments of the new astronomy.

The astronomers, tourists, and frontier residents who gathered beneath the clear skies of the Rockies that summer afternoon witnessed more than America's first "great eclipse." Whether they knew it or not, they were present at a turning point in American science. Astronomy was headed in a new direction.

Upward, into the mountains.

FURTHER READING AND NOTES

Thank you for reading *America's First Great Eclipse*. To learn more, including where to find information on the upcoming 2017 eclipse (and those to follow), please visit FirstGreatEclipse.com.

* * *

The 1878 eclipse is not as well-known as it deserves to be, which is why I wrote this book. One early article on the subject is "The Great Eclipse of 1878," by John A. Eddy, in *Sky and Telescope* (June 1973), 340–346. Eddy also wrote an excellent account of Edison and the tasimeter in Wyoming: "Thomas A. Edison and Infra-red Astronomy," *Journal for the History of*

Astronomy (1972), 165–187. For the role of the US Naval Observatory in funding and organizing the 1878 eclipse expeditions, see Steven J. Dick's, *Sky and Ocean Joined: The U.S. Naval Observatory 1830–2000* (2003). On nineteenth-century eclipse expeditions in general, including the 1878 eclipse, Alex Pang's *Empire and the Sun: Victorian Solar Eclipse Expeditions* (2002) explores interesting interactions between astronomy and culture, and professional and amateur eclipse observers. Another book on the eclipse, *American Eclipse* by David Barron, is scheduled for publication after this one, so I have not read it, but it looks to be quite good, as well.

Readers may be interested in reviewing the US Naval Observatory's official account of the 1878 eclipse expeditions: *Reports on the Total Solar Eclipses of July 29, 1878, and January 11, 1880* (Washington: Government Printing Office, 1880). Also of interest is the US Naval Observatory's thirty-page guide to observing the eclipse, *Instructions for Observing the Total Solar Eclipse of July 29, 1878, Issued by the U.S. Naval Observatory* (Washington: Government Printing Office, 1878). Finally, Samuel Pierpont Langley discusses his experience on Pikes Peak in his book *The New Astronomy* (1888).

On the search for Vulcan, two excellent and highly readable accounts are Richard Baum and William Sheehan's *In Search of Planet Vulcan: The Ghost in Newton's Clockwork Universe* (1997; see especially chapter fourteen, "The Planet Hunters Come to Indian Country"), and Thomas Levenson's *The Hunt for Vulcan* (2015).

Finally, on the general history and cultural impact of eclipses, there are a number of excellent books. The most recent is *Sun Moon Earth* by astronomer Tyler Nordgren (2016), which also discusses the hunt for Vulcan. I also like *Totality: Eclipses of the Sun* (2008) by Mark Littmann, et al.

* * *

A Note on Quotations and Spellings

In this book, the quotations I've used are mostly repeated verbatim. But I have occasionally altered punctuation or made other edits for readability, such as omitting ellipsis when shortening quotations, or adding clarifying words, while being careful not to change the meaning of the original.

Also, whenever possible but without sacrificing clarity, I retained the spellings of words from the 1870s. For example, today "Pikes Peak" is spelled without the possessive apostrophe. But in 1878 most people still spelled it "Pike's Peak," with an apostrophe. So I've retained their spelling of that and many other words in quotations, unless doing so was simply too distracting to the reader.

Acknowledgments and Sources

As an academic who normally writes for other academics, it went against years of training to not fill this short book with endless footnotes, endnotes, and other such scholarly decorations. But this is a book written for a popular audience, and so I left them out. However, I did want to list some of the people on whose expertise I relied, and list some of the main sources I used as I wrote this book.

First, I would like to thank the following: Steven J. Dick, former NASA Chief Historian, whose expertise is unrivaled; John Briggs, astronomer and past president of the Antique Telescope Society, who answered my questions about 1870s-era telescopes and solar filters; Barbara Becker, who helped me get a better grasp of nineteenth-century spectroscopy; and David DeVorkin, Senior Curator at the Smithsonian National Air and Space Museum, who pointed me toward a number of important sources many years ago.

I would also like to thank the staff of the following archives and institutions for their help: the Colorado Springs Pioneers Museum; the Colorado Historical Society; the Pikes Peak Library District Special Collections; the United States Naval Observatory Library; the Carbon County Museum, Wyoming; and the Vassar College Library.

Audience members at the following conferences offered many useful comments, which helped me rethink and revise my narrative: the Eighth Notre Dame Biennial History of Astronomy Workshop (Notre Dame, Indiana, 2007); the Twenty-Third International Congress of History of Science and Technology (Budapest, Hungary, 2009); and the Western History Association Fifty-Second Annual Conference (Denver, Colorado, 2012).

A few readers of early drafts of this book also offered helpful comments. These include Bill Cherf, Stella Cottam, Sean Golden, Martin Shoemaker, and Greg Vose. Finally, my editor, Shelley Holloway, did her usual amazing job.

Although I have presented and published some of the information in this book before, and have received helpful feedback from many people (for which I am extremely grateful), any errors it contains are my own.

* * *

I consulted many books, journal articles, newspaper stories, diaries, and other materials while preparing this book. What follows is a select bibliography of the most significant of those sources.

Barker, George F. "On the Total Solar Eclipse of July 29th, 1878." *Proceedings of the American Philosophical Society* 18, no. 102 (1878): 103-14.

Baum, Richard and William Sheehan. *In Search of Planet Vulcan: The Ghost in Newton's Clockwork Universe.* New York: Plenum Publishing Corporation, 1997.

Becker, Barbara. Unravelling Starlight: William and Margaret Huggins and the Rise of the New Astronomy. New York: Cambridge University Press, 2011.

Bell, Trudy E. "The Victorian Space Program," *The Bent of Tau Beta Pi* (Spring 2003), 11–18.

Clerke, Agnes M. A Popular History of Astronomy During the Nineteenth Century. London: Adam and Charles Black, 1902.

Cottam, Stella and Wayne Orchiston. Eclipses, Transits, and Comets of the Nineteenth Century: How America's Perception of the Skies Changed. New York: Springer, 2015.

Crowe, Michael J. Modern Theories of the Universe from Herschel to Hubble. New York: Dover, 1994.

Dick, Steven J. *Sky and Ocean Joined: The U.S. Naval Observatory, 1830–2000.* New York: Cambridge University Press, 2003.

Eddy, John A. "The Great Eclipse of 1878," *Sky and Telescope* (June 1973), 340–346.

Eddy, John A. "Thomas A. Edison and Infra-red Astronomy." *Journal for the History of Astronomy* (1972), 165–187.

Holden, Edward S. *Mountain Observatories in America and Europe*. Washington, DC: Smithsonian Institution, 1896.

Langley, Samuel P. *The New Astronomy*. Boston: Ticknor and Company, 1888.

Levenson, Thomas. The Hunt for Vulcan …and How Albert Einstein Destroyed a Planet, Discovered Relativity, and Deciphered the Universe. Penguin Random House, 2016.

Lockyer, J. Norman. *The Chemistry of the Sun.* New York: Macmillan & Co., 1887.

Mitchell, Maria. *Life, Letters and Journals*. Compiled by Phoebe Mitchell Kendall. Boston: Lee and Shepard Publishers, 1896.

Newton, Isaac. *Opticks.* New York: Dover, 1979.

Nordgren, Tyler. Sun Moon Earth: The History of Solar Eclipses from Omens of Doom to Einstein and Exoplanets. New York: Basic Books, 2016.

Pickering, Edward C. "Mountain Observatories." *Appalachia*, 1884.

Reports on the Total Solar Eclipses of July 29, 1878, and January 11, 1880. Washington DC: Government Printing Office, 1880.

Ruskin, Steve. "Advancing Astronomy on the American Frontier: The Career of Frank Herbert

Loud." *Journal of Astronomical History and Heritage*, 2012.

Ruskin, Steve. "'Among the Favored Mortals of Earth': The Press, State Pride, and the Great Eclipse of 1878." *Colorado Heritage*, 2008.

Scott, Hugh Lenox. *Some Memories of a Soldier*. New York: The Century Company, 1928.

Smyth, Charles Piazzi. Teneriff, An Astronomer's Experiment: Or, Specialties of a Residence Above the Clouds. London: Lovell Reeve, 1858.

Soojung-Kim Pang, Alex. *Empire and the Sun: Victorian Solar Eclipse Expeditions*. Stanford: Stanford University Press, 2002.

"The Eclipse: A Grand Spectacle." *Colorado Springs Weekly Gazette*, August 3, 1878.

"The Eclipse! A Glorious Day for Colorado!" *Georgetown Colorado Miner*, August 3, 1878.

"The Eclipse of the Sun: Successful Observations in the West." *New York Times*, July 30, 1878.

"The Eclipse of the Sun." *Nature*. August 22, 1878, 430–434.

"The Great Eclipse." *Laramie Daily Sentinel*, July 30, 1878.

"The Great Eclipse." *Laramie Weekly Sentinel*, August 3, 1878.

"The Great Solar Eclipse," *Harper's Weekly: A Journal of Civilization*, August 24, 1878.

Todd, Mabel Loomis. *Total Eclipses of the Sun*. Boston: Roberts Brothers, 1894.

Young, Charles Augustus. *The Sun.* New York: D. Appleton and Company, 1881.

About the Author

Steve Ruskin is an award-winning historian of astronomy, with a PhD in History and Philosophy of Science from the University of Notre Dame. He is the author of one previous book and over fifty articles, chapters, and reviews. He was a visiting researcher at Cambridge University, England, on a grant from the National Science Foundation, and is an alumnus of the Launch Pad Astronomy Workshop. He currently serves as the moderator of HASTRO-L, the long-running history of astronomy listserv, and is on the Board of Advisors for the National Space Science & Technology Institute. He is a native of Colorado Springs, Colorado.

Made in the USA
Lexington, KY
22 June 2017